T0331673

ADVANCED STATISTICAL MECHANICS

Recommended Titles in Related Topics

Lectures on Statistical Mechanics
by Berthold-Georg Englert
ISBN: 978-981-122-457-7
ISBN: 978-981-122-554-3 (pbk)

Thermodynamics and Statistical Mechanics
by Richard Fitzpatrick
ISBN: 978-981-122-335-8
ISBN: 978-981-122-423-2 (pbk)

Introduction to Statistical Mechanics
by John Dirk Walecka
ISBN: 978-981-4366-20-5
ISBN: 978-981-4366-21-2 (pbk)

ADVANCED STATISTICAL MECHANICS

JIAN-SHENG WANG

NATIONAL UNIVERSITY OF SINGAPORE, SINGAPORE

World Scientific

EW JERSEY · LONDON · SINGAPORE · BEIJING · SHANGHAI · HONG KONG · TAIPEI · CHENNAI · TOKYO

Published by

World Scientific Publishing Co. Pte. Ltd.
5 Toh Tuck Link, Singapore 596224
USA office: 27 Warren Street, Suite 401-402, Hackensack, NJ 07601
UK office: 57 Shelton Street, Covent Garden, London WC2H 9HE

British Library Cataloguing-in-Publication Data
A catalogue record for this book is available from the British Library.

ISBN 978-981-124-214-4 (hardcover)
ISBN 978-981-124-215-1 (ebook for institutions)
ISBN 978-981-124-216-8 (ebook for individuals)

For any available supplementary material, please visit
https://www.worldscientific.com/worldscibooks/10.1142/12414#t=suppl

Typeset by Stallion Press
Email: enquiries@stallionpress.com

Printed in Singapore

To Esther, Francis, and Aloysius

Preface

Statistical mechanics is a fundamental subject that every graduate student of physics needs to learn. It is fundamental as physical systems cannot exist in isolation, either because of small but definite interactions with other parts of the world or interactions among similar entities, such as a collection of gas molecules. This book arises out of many years of teaching of a course known as "Advanced Statistical Mechanics." This book, parallel to the taught course, will begin by reviewing the basics of equilibrium statistical mechanics. The materials should already have been learned in an undergraduate statistical mechanics class. Thus, the coverage will be somewhat formal and brief. We will then go on to interacting systems, such as the van der Waals gas and the Ising model, with mean-field theories of phase transitions, and certain exact methods. The second part of the book will contain more advanced topics of nonequilibrium statistical mechanics, which covers Brownian motion, Langevin equations, Fokker-Planck equations, Kubo's linear response theory, Boltzmann equations, and the Jarzynski equality. The book, as a whole, offers a semester's worth of contents — about 13 weeks of lectures. I have also collected a good number of problems which should be helpful to students in reinforcing fundamental concepts.

I like to take this opportunity to thank the Department of Physics of National University of Singapore for giving me the privilege to teach this course. A good deal of interactions over the years with the students taking the course sharpened my own understanding of the subject. I also thank Professor Robert H. Swendsen for introducing me to the world of statistical mechanics and for exchanges over the concept of entropy lately. Finally, I thank my family members for their love and continued support.

Contents

9.   Brownian Motion — Langevin and Fokker-Planck Equations   133

10. Systems Near and Far from Equilibrium — Linear Response Theory and Jarzynski Equality   161

11. The Boltzmann Equation   177

Chapter 1

Thermodynamics

1.1 Introduction

Thermodynamics is formulated out of our understanding of heat, from the early 1800s to the end of the nineteenth century. Its foundations include the works of Mayer and Joule for the equivalence of heat and mechanical work, and that of Carnot and Clausius related to the concept of entropy. It is just as useful today as it was then due to its fundamental nature and great generality.

Statistical mechanics tries to give a microscopic interpretation and explanation of thermodynamics, and much more. For this reason, to understand statistical mechanics, it is beneficial to have a brief overview of thermodynamics. There are at least three ways of presenting thermodynamics. The first is historical [Fermi (1937); Pippard (1957)], which we will quickly mention now. This formulation starts with the first law of thermodynamics, which states that internal energy U is a state function, and energy is conserved, with the equation $U_2 - U_1 = \Delta U = Q + W$, where Q is heat absorbed by the system and W is mechanical work done to the system, for a process going from state 1 to state 2. A key point here is that while the heat Q and work W depend on the details of the process, the sum does not. Then we introduce the second law of thermodynamics, through the Carnot cycle of heat engine, obtaining the inequality for the entropy change, $\Delta S \geq Q/T$. Here T is the absolute temperature. And lastly, the third law of thermodynamics gives a meaning to the absolute magnitude $S = 0$ when temperature T approaches 0.

The second, alternative modern formulation is that by Callen [1985]. This formulation is in terms of postulates instead of laws. Callen's formulation of thermodynamics is convenient for making a quick connection to

1

statistical mechanics. This connection is through the so-called Boltzmann principle,

$$S = k_B \ln \Omega, \qquad (1.1)$$

where k_B is the Boltzmann constant (now defined as exactly $1.380\,649 \times 10^{-23}\,\mathrm{J/K}$), and Ω is the number of states that is consistent with a given set of macroscopic constraints, one of which is a fixed energy U. The Callen formulation of thermodynamics focuses on entropy S. Callen postulated four axioms that entropy must obey. From these postulates, all of thermodynamics is derived. In the next section, we will give some details of this formulation.

A third, "post-modern" formulation, due to Lieb and Yngvason, exists. It is a rigorous mathematical formulation of thermodynamics which we will not discuss here. Interested readers can consult the literature [Lieb and Yngvason (1999)].

1.2 Callen's postulates

To begin with, we need to define states, or more precisely, equilibrium states. Thermodynamics is intended to work only for or between equilibrium states. Nonequilibrium states cannot be rigorously described by thermodynamics. An equilibrium state, at least, cannot be time dependent. But this is not enough. In addition to being time independent for all macroscopic physical observables, there cannot be transport of any kind, such as of charge, or energy. This rules out nonequilibrium steady states, which, although time independent, may have transport phenomenon taking place. Rigorously speaking, equilibrium states will be these that satisfy the postulates. This is very much like the concept of a "point" in Euclidean geometry, which is an undefined term but fixed by the postulates of the geometry. For simplicity of discussion, we assume that the systems in equilibrium states are homogeneous, isotropic, uncharged, and without surface effects. Good examples of such systems are a bottle of gas, a cup of water, or a piece of steel, under normal conditions. These will be termed *simple systems*.

The first postulate is about the existence of equilibrium states:

Postulate I. *"There exist particular states (called equilibrium states) of simple systems that, macroscopically, are characterized completely by the internal energy U, the volume V, and the number of particles N."*

The energy U is fundamental in thermodynamics, as can be seen from its role in the first law of thermodynamics. All thermodynamic systems should have such a variable to characterize an equilibrium state. For an ideal gas, the volume V and particle number N together with U completely specify the state. However, for some other systems, other macroscopic variables may be needed, and the volume may be an irrelevant parameter. For example, for a magnetic system, we need the total magnetization as an additional variable. The point here is not about the choice of variables, but the fact that an equilibrium state is characterized by only a small number of variables, while a nonequilibrium system, such as a fluid flow, requires a nearly infinite number of variables to specify its states.

A second important concept is the composite system. A composite system is just a few simple systems (say two), placed together, without interactions among them. For example, an ice cube maintained at 0 degree Celsius and a cup of coffee at 96 degree Celsius without making contact with each other is a composite system. This concept of composite systems is needed to state postulates II and III below.

Postulate II. *"There exists a function (called the entropy S) of the extensive parameters of any composite system, defined for all (constrained) equilibrium states and having the following property: the values assumed by the extensive parameters in the absence of an internal constraint are those that maximize the entropy over the manifold of constrained equilibrium states."*

Extensive parameters are parameters that scale with the system — they double when the system is doubled. For simple subsystems of gases, they are U_i, V_i, N_i, $i = 1, 2, \ldots$. Intensive parameters are independent of the size of the system, such as temperature or pressure. The second postulate is a principle of maximum entropy. However, unlike Boltzmann's original approach based on a formula for entropy, we don't have a formula here. All we know is that such a quantity exists and is defined only for constrained equilibrium states. To better understand the maximal principle, we need the next postulate.

Postulate III. *"The entropy of a composite system is additive over the constituent subsystems. The entropy is continuous and differentiable and is a monotonically increasing function of the energy."*

The additivity of entropy gives us a way to compute the entropy of a system consisting of two subsystems, by $S = S_1 + S_2$. By postulate I, S_1 is a function of U_1, V_1, N_1, and similarly, S_2 is a function of U_2, V_2, and N_2. These values stay constant because of the internal and external constraints — each subsystem is isolated from the outside world. They are in fixed containers, unable to change volume or exchange particles or energy.

Imagine that we take these isolated boxes and connect them with a heat conducting metal wire or a metal wall. Internal energy can then be transferred between them, with the total energy $U = U_1 + U_2$ remaining conserved. This partition of energy is an internal constraint. One of the most important problems in thermodynamics is to determine the equilibrium state that eventually results after the removal of internal constraints in a closed, composite system. Postulate II answers the question: if the constraint is removed, allowing free transfer of energy between them, what the values of U_i are when the new overall equilibrium is reached? It is that when $S = S_1(U_1, V_1, N_1) + S_2(U - U_1, V_2, N_2)$ is maximized with respect to U_1.

Constraints are usually implemented by "walls." An adiabatic wall cannot transfer heat, a diathermal wall can transfer heat but not matter. A movable wall can do work, a semi-permeable wall can let certain type of molecules pass through.

Postulate IV. "*The entropy of any system vanishes in the state for which* $T \equiv (\partial U / \partial S)_{V,N} = 0$ *(that is, when the absolute temperature is zero).*"

The last postulate is equivalent to the usual third law of thermodynamics. Note that graphically (see Fig. 1.1), $S(U)$ comes down vertically with

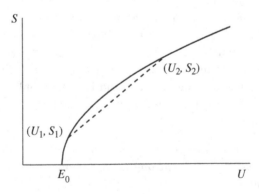

Fig. 1.1 Entropy as a function of internal energy for a Fermi liquid, $S \propto \sqrt{U - E_0}$. Note that S approaches 0 vertically with an infinite slope at the ground state energy E_0. Concavity of entropy means that the value of S is not below (in a segment) the dotted straight line for any choices of the end points.

an infinite slope at $S = 0$. Nothing is said about the energy U when that happens. Since mechanics and thermodynamics remain invariant when the energy scale is changed from U to $U +$ const, $U = 0$ has no significance.

We have drawn the figure in 1.1 so that S is a monotonically increasing function of U. Since we will define the slope of this curve as $1/T$, postulate III implies that the temperature is always a positive quantity. However, there are physical systems (such as the Ising model) for which entropy decreases after passing a maximum. For such systems, negative temperature is allowed, and Callen's postulate III needs to be revised.

1.3 Definition of temperature

Since postulate III assumes that entropy is a monotonically increasing function of energy, it is convenient to invert the relation, to write $U = U(S, V, N)$. Such a functional relationship is known as a fundamental relation. The fundamental relation of a particular system contains the complete thermodynamic information about the system. The differential of the energy function is

$$dU = \left(\frac{\partial U}{\partial S}\right)_{V,N} dS + \left(\frac{\partial U}{\partial V}\right)_{S,N} dV + \left(\frac{\partial U}{\partial N}\right)_{S,V} dN$$

$$= T\,dS - p\,dV + \mu\,dN. \tag{1.2}$$

Here, the second line is to be interpreted as definitions for thermodynamic temperature, pressure, and chemical potential, respectively:

$$\left(\frac{\partial U}{\partial S}\right)_{V,N} \equiv T, \quad \left(\frac{\partial U}{\partial V}\right)_{S,N} \equiv -p, \quad \left(\frac{\partial U}{\partial N}\right)_{S,V} \equiv \mu. \tag{1.3}$$

The subscripts under the parentheses indicate explicitly what variables are fixed when the partial derivative is taken, an important feature in thermodynamics. For multi-component systems with several chemical species, such as a water and methanol mixture, each component has an associated chemical potential. Thus the last term in Eq. (1.2) could be a sum, $\sum_i \mu_i dN_i$. For simplicity of presentation, in this book, we will almost exclusively consider systems of one chemical species.

The differentials, for example, dU, mean we are comparing the physical quantities of the same system at two infinitesimally close states, each in thermal equilibrium. We are not concerned with the system evolving from an initial to a final state. Now we define heat $\delta Q = TdS$ and work $\delta W = -pdV$. The fundamental thermodynamic relation then becomes

$dU = \delta Q + \delta W + \mu dN$. This is just the first law of thermodynamics. Here, the special symbol δ is used for work and heat, as they depend on how the process goes (they are path dependent), while dU is a total differential which depends only on the end points.

There is some subtlety associated with $\delta Q = TdS$. This is only strictly correct for a (reversible) process that is nearly in equilibrium at all times even as the system is changing. In general, we should have the Clausius inequality,

$$\delta Q \leq TdS, \tag{1.4}$$

which is another way of stating the second law of thermodynamics.

Back to the definition of temperature. Since

$$\left(\frac{\partial U}{\partial S}\right)_{V,N} \left(\frac{\partial S}{\partial U}\right)_{V,N} = 1,$$

we also have

$$\left(\frac{\partial S}{\partial U}\right)_{V,N} \equiv \frac{1}{T}. \tag{1.5}$$

A question then arises as to whether such a definition of temperature is in agreement with our usual understanding of temperature [Maxwell (1888)]. For instance, when two bodies are in thermal contact and in equilibrium, are temperatures of the two bodies the same? or does heat flow from high to low temperatures? By applying Callen's postulate II, we can demonstrate that the answer is yes on both counts.

Consider two subsystems 1 and 2, fixing other parameters and focusing only on the energies, with the total energy being conserved, $U = U_1 + U_2 =$ const, and total entropy being additive, $S = S_1 + S_2$. We can determine the equilibrium condition as postulated in II. When the entropy reaches a maximum, its differential must be 0,

$$dS = \left(\frac{\partial S_1}{\partial U_1}\right)_{V_1,N_1} dU_1 + \left(\frac{\partial S_2}{\partial U_2}\right)_{V_2,N_2} dU_2$$
$$= \left(\frac{1}{T_1} - \frac{1}{T_2}\right) dU_1 = 0. \tag{1.6}$$

We have used the fact $dU_1 + dU_2 = 0$. Thus, the equilibrium condition implies that $1/T_1 = 1/T_2$, or the condition of thermal equilibrium is that the temperatures must be the same, $T_1 = T_2$.

Then, does heat flow from hot to cold? Let us assume that $T_1 > T_2$. Since entropy must increase from the initially separated systems to the final equilibrium state, we have $dS > 0$. The above equation, before setting it to 0, implies $dU_1 < 0$ (since $1/T_1 - 1/T_2 < 0$). This shows that system 1 loses energy, i.e., heat flows from hot to cold, which is the Clausius statement of the second law of thermodynamics.

We will now discuss the units used for the fundamental thermodynamic equation, $dU = TdS - pdV + \mu dN$. Energy in SI units is joule, and temperature is measured in kelvin, having the symbol K. Entropy has the units joule/kelvin. The Kelvin temperature scale is fixed by energy through the exactly defined Boltzmann constant such that T kelvin corresponds to $k_B T$ joule. An older definition of the scale with the absolute zero, $T = 0$, the minimal possible temperature fixed by nature and the triple point of water, where water, ice, and water vapor can coexist, defined as $T_t = 273.16\,\mathrm{K}$, is considered to be obsolete. The daily usage of the Celsius scale is related to the Kelvin scale by $t(^\circ\mathrm{C}) = T(\mathrm{K}) - 273.15$, while the Fahrenheit scale is related to the Celsius scale by $T(^\circ\mathrm{F}) = \frac{9}{5}t(^\circ\mathrm{C}) + 32$.

1.4 Some formal thermodynamic relations

We discuss some of the useful thermodynamic relations as a consequence of the postulates.

(A) The additivity of entropy together with the fact that U, V, N are extensive variables implies

$$U(\lambda S, \lambda V, \lambda N) = \lambda U(S, V, N), \tag{1.7}$$

that is, U as a function of S, V, N is a homogeneous function of degree one. differentiating with respect to λ on both sides, and then setting $\lambda = 1$, we obtain

$$TS - pV + \mu N = U. \tag{1.8}$$

This is known as the Euler equation.

(B) Taking the differential of the Euler equation, and subtracting from it the fundamental equation, we obtain

$$dU = TdS + SdT - pdV - Vdp + \mu dN + Nd\mu,$$
$$0 = SdT - Vdp + Nd\mu. \tag{1.9}$$

The second equation is known as the Gibbs-Duhem relation. The Gibbs-Duhem relation is valid for macroscopically large systems where surface effects are ignored.

(C) Mixed partial derivatives are equal (provided that the second derivatives exist and are continuous), for example,

$$\frac{\partial^2 U}{\partial S \partial V} = \frac{\partial^2 U}{\partial V \partial S}. \tag{1.10}$$

This gives us

$$-\left(\frac{\partial p}{\partial S}\right)_{V,N} = \left(\frac{\partial T}{\partial V}\right)_{S,N}. \tag{1.11}$$

There are many relations of this kind, and they are known as Maxwell's relations.

(D) Heat capacity is an experimentally measurable quantity that partially characterizes a thermodynamic system. It is defined by $C = \delta Q/dT$. The amount of heat absorbed depends on the specific process. The heat capacity at constant volume is

$$C_V \equiv \left(\frac{\delta Q}{dT}\right)_{V,N} = T\left(\frac{\partial S}{\partial T}\right)_{V,N} = \left(\frac{\partial U}{\partial T}\right)_{V,N}. \tag{1.12}$$

We can show that the heat capacity at constant volume must be a positive quantity. This is due to the concavity of entropy, implied by the postulates:

$$\lambda S(U_1) + (1 - \lambda)S(U_2) \leq S\big(\lambda U_1 + (1 - \lambda)U_2\big), \quad 0 \leq \lambda \leq 1. \tag{1.13}$$

A geometric meaning of this equation is that the curve $S(U)$ always lays above the straight line joining the points (U_1, S_1) and (U_2, S_2), see Fig. 1.1. A local form of this inequality is

$$\left(\frac{\partial^2 S}{\partial U^2}\right)_{V,N} \leq 0, \tag{1.14}$$

which can be rewritten as

$$\frac{\partial^2 S}{\partial U^2} = \frac{\partial}{\partial U}\left(\frac{\partial S}{\partial U}\right) = \frac{\partial}{\partial U}\left(\frac{1}{T}\right)$$

$$= -\frac{1}{T^2}\frac{\partial T}{\partial U} = -\frac{1}{T^2 C_V} \leq 0. \tag{1.15}$$

Thus, $C_V \geq 0$.

1.5 Supplementary reading — differential, total differential, and line integral

In this supplementary section, we recall some mathematical notions about differentials and their integrals. Given a function of a single variable, $f(x)$, its derivative at point x is

$$f'(x) = \lim_{\Delta x \to 0} \frac{f(x + \Delta x) - f(x)}{\Delta x}. \tag{1.16}$$

A differential of f is a linear function of dx, $df = f' dx$. Here dx is a change in the independent variable x that is supposed to be (infinitesimally) small. The idea of differentials can be generalized to multiple variables. For example, for a function of two variables, $f(x, y)$,

$$df = \frac{\partial f}{\partial x} dx + \frac{\partial f}{\partial y} dy. \tag{1.17}$$

The partial derivatives are evaluated at the point (x, y). The partial derivatives form a two-dimensional vector, $(\partial f / \partial x, \partial f / \partial y)$, known as the gradient.

Now we ask the reverse question. If we are given a differential of the form $g(x, y)dx + h(x, y)dy$, is there a function $f(x, y)$ such that $df = g dx + h dy$? This is equivalent to asking if there exists a function f, such that its partial derivatives with respect to x and y are precisely g and h. If the answer to this question is positive, we call the differential a 'total differential', or 'exact'. The condition for the existence of a total differential is $\partial h / \partial x - \partial g / \partial y = 0$. This can be seen as the condition that the mixed partial derivatives of f must be equal. More general statements can be given, even in higher dimensions, if one studies the theory of differential forms [Rubin (1976)].

We can define line integral on the (x, y)-plane, as

$$\int_C \left(g \, dx + h \, dy \right). \tag{1.18}$$

Here C specifies a path on the plane, from point 1 to point 2. (dx, dy) is an infinitesimal vector connecting two nearby points on the line. If the integrand is a total differential, then the value of the integral depends only on the end points, giving the Newton-Leibniz result $f_2 - f_1$. However, if the integrand is not a total differential, such as the work done on a path, then one has to perform the line integral and the result depends explicitly on the path taken.

Problems

Problem 1.1. (a) What is a quasi-static process? (b) What is a triple point? How is the Kelvin SI scale defined? (c) What is the definition of (electro-) chemical potential? (d) Why do we need to postulate that $\partial S/\partial U \geq 0$? (e) Why do we need $\partial^2 S/\partial U^2 \leq 0$? (f) Under what conditions does entropy increase? (g) State the basic/fundamental properties of temperature T. (h) State the fundamental properties of entropy S. (i) What is an adiabatic wall, a diathermal wall, a semi-permeable wall, a closed system, an open system, and an isolated system?

Problem 1.2. Two pieces of information are needed in order to determine the entropy of an ideal gas, the equation of state, $pV = Nk_BT$, and the calorimetric information, $C_V = \frac{3}{2}Nk_B$. Determine the entropy $S = S(U, V, N)$ using thermodynamic relations.

Problem 1.3. An adiabatic process is defined by $\delta Q = 0$. Consider the quasi-static, adiabatic process of an ideal gas of N particles, starting from the state (U_1, V_1), determine the change of internal energy $U_2 - U_1$, change of temperature $T_2 - T_1$, work done to the system W, and heat absorbed Q, when the volume of the system is expanded to twice the original, $V_2 = 2V_1$.

Problem 1.4. Thermal radiation or photons can be described by equilibrium thermodynamics. The pressure due to thermal photons is given by $p = \frac{1}{3}U/V$, where U is internal energy, and V is volume. It is also known that the energy density $u = U/V$ is a function of the temperature T only. Particle number N is not a state variable as photon numbers are not a conserved quantity. Based on the information given, derive the entropy $S(U, V)$ of thermal radiation. [Note the basic thermodynamic equation $dU = TdS - pdV$ and the Maxwell relation due to $\partial^2 S/\partial U \partial V = \partial^2 S/\partial V \partial U$].

Problem 1.5. The entropy S of a hypothetical system as a function of internal energy U has the following form:

$$S = \begin{cases} a\sqrt{UN}, & U < U_1, \\ bU, & U_1 \leq U \leq U_2, \\ a'\sqrt{UN} + cU, & U_2 < U. \end{cases}$$

Here N is the number of particles; a, a', b, and c are some positive constants. It turns out that these constants cannot be chosen arbitrarily but are uniquely determined by a', U_1 and U_2. (a) Draw a qualitative plot of S as *a function of U.* (*b*) *Compute temperature T as a function of U. Draw a qualitative plot of U as a function of temperature T.* (*c*) *Write down a set of four equations determining the parameters a, b, and c. Solution is not required.* (*d*) *The straight line segment in S vs. U represents a first-order phase transition. Determine the latent heat of the phase transition.* (*e*) *The straight line segment also represents a coexistence of two phases. Explain briefly?*

Problem 1.6. (*a*) *Consider three variables x, y, and z which are related functionally, i.e., x = x(y, z), or y = y(x, z), or z = z(x, y), show that*

$$\left(\frac{\partial x}{\partial y}\right)_z \left(\frac{\partial y}{\partial z}\right)_x \left(\frac{\partial z}{\partial x}\right)_y = -1.$$

(*b*) *Using this relation to show that the thermal expansion coefficient α can be written in two alternative forms,*

$$\alpha \equiv \frac{1}{V}\left(\frac{\partial V}{\partial T}\right)_p = \frac{1}{B}\left(\frac{\partial p}{\partial T}\right)_V,$$

where B is called the bulk modulus. Work out the definition of bulk modulus B.

Problem 1.7. *Using the thermodynamic relations, show that the heat capacity at constant pressure is related to the heat capacity at constant volume by*

$$C_p = C_V + \frac{TV\alpha^2}{k_T},$$

where $\alpha = V^{-1}(\partial V/\partial T)_p$ is the thermal expansion coefficient and $\kappa_T = -V^{-1}(\partial V/\partial p)_T$ is the isothermal compressibility.

Chapter 2

Foundation of Statistical Mechanics, Statistical Ensembles

In this chapter, we introduce the basic assumptions of equilibrium statistical mechanics and discuss the starting point and important formulas for calculations. We first discuss the origin of the microcanonical ensemble and apply this fundamental ensemble to the determination of the entropy of a classical ideal gas. We then derive the canonical ensemble and show that the partition function is related to the Helmholtz free energy. We end the chapter with some simple applications of non-interacting oscillators and illustrate the conditions for ensemble equivalence.

2.1 Classical dynamics

The Newtonian classical world view is deterministic. Once the initial positions and velocities of a mechanical system are known, all future developments can be predicted. Newton's equation of motion is $\mathbf{F} = m\mathbf{a}$. This equation is easily generalized for a collection of particles, each following Newton's equation and with some form of interacting forces between them. Alternatively, in analytic mechanics, we can also use the Lagrange formulation, with Hamilton's principle of least action, $\delta \int L\, dt = 0$. This gives the Euler-Lagrange equation of motion. However, in equilibrium statistical mechanics, the most relevant formulation of mechanics is that of Hamiltonian dynamics. This has to do with the fact that phase space volume Ω is a conserved quantity. In Hamiltonian dynamics [Goldstein *et al.* (2002)], the system is defined by a Hamiltonian function of the generalized coordinates q_j and their conjugate momenta p_j, which is related to the

Lagrangian function by

$$H(q,p) = \sum_j \dot{q}_j p_j - L(q,\dot{q}). \tag{2.1}$$

To be qualified as a Hamiltonian, the velocity variables have to be replaced by the momentum variables through the implicit equations $p_j = \partial L/\partial \dot{q}_j$. The Hamilton equations of motion are

$$\dot{q}_j = \frac{\partial H}{\partial p_j}, \qquad \dot{p}_j = -\frac{\partial H}{\partial q_j}, \qquad j = 1, 2, \ldots. \tag{2.2}$$

For a harmonic oscillator the Hamiltonian can be written as $H(x,p) = p^2/(2m) + kx^2/2$. From this we get $\dot{x} = p/m$ and $\dot{p} = -kx$, or $m\ddot{x} + kx = 0$, which is Newton's equation of motion for a harmonic oscillator.

The phase space is the set of points $\Gamma = (q,p) = (q_1, q_2, \ldots, p_1, p_2, \ldots)$ of all coordinates and momenta. It is an even dimensional manifold, 2, 4, $\ldots$. For notational brevity, instead of listing all the coordinates and momenta, we will use the more abstract notation Γ to denote a point in phase space. Once a point is specified at a given time t, its future locations and also past locations are determined by solving the Hamilton equations of motion. In fact, any function A of phase space point Γ is also determined. This is given by the equation

$$\dot{A} = (A, H). \tag{2.3}$$

Here we assume that the time-dependence in A is only through Γ. The parentheses denote the classical Poisson bracket:

$$(A, B) = \sum_j \left(\frac{\partial A}{\partial q_j} \frac{\partial B}{\partial p_j} - \frac{\partial B}{\partial q_j} \frac{\partial A}{\partial p_j} \right). \tag{2.4}$$

For example, $(q_j, p_k) = \delta_{jk}$. With this definition, we can see that Eq. (2.3) is the same as the Hamilton equation for q or p if we take A to be q_j or p_j. We can also solve Eq. (2.3) formally as [Balescu (1975)]

$$A_t \equiv A(\Gamma_t) = A(\Gamma, t) = e^{t(\cdot, H)} A(\Gamma, 0). \tag{2.5}$$

The meaning of this equation needs elaborating. We should take Γ as the initial value at time $t = 0$, and the equation predicts what the value of A will be at a later time t. This is obtained by a Taylor expansion of $A(\Gamma, t) = A(\Gamma, 0) + t dA/dt|_{t=0} + \cdots$. The notation $(\cdot, H)$ indicates a differential operator such that $(\cdot, H)f \equiv (f, H)$. The action of $e^{t(\cdot, H)}$ to A is understood by an expansion of the operator exponential.

In a more advanced treatment of a Hamiltonian dynamical system, its features and properties can be understood 'geometrically'. An important geometric feature of the Hamiltonian system is its symplectic structure [Arnold (1989)], characterized by the differential 2-form:

$$\omega^2 = \sum_j dq_j \wedge dp_j, \tag{2.6}$$

where $\wedge$ denotes the antisymmetric wedge product (for example, $dq_1 \wedge dq_1 = 0$ and $dq_1 \wedge dq_2 = -dq_2 \wedge dq_1$). This quantity (viewed as denoting an oriented surface area element) and its power $(\omega^2)^k$, for $k = 1, 2, \ldots, N$, are invariant with respect to the dynamics, in the sense that, for a fixed t, the canonical transformation, $\Gamma \rightarrow \Gamma_t$, preserves the area, 4-dimensional surfaces, $\ldots$, and finally the phase space volume $d\Gamma = (\omega^2)^N / N!$ for a system of N degrees of freedom. The last of these is the statement of Liouville's theorem.

2.2 Statistical description

For systems that consist of a large number of degrees of freedom, we are not interested in the actual trajectories Γ_t as a function of time t. This contains too much information. We prefer a simpler description with a few numbers. In particular, thermodynamics describes a macroscopic system with energy U, volume V, particle number N, entropy S, pressure p, etc., which can be obtained either as a constraint, or an average over a large number of particles. For this reason, a "statistical" description is a useful and fruitful approach.

Fundamental to the statistical description is the concept of probability density ρ in phase space, where $\rho(\Gamma, t)d\Gamma$ gives the probability to find the system near point $\Gamma = (q_1, q_2, \ldots, p_1, p_2, \ldots)$ at time t in volume element $d\Gamma = dq_1 dq_2 \cdots dp_1 dp_2 \cdots$. Since $\rho d\Gamma$ represents probability, we have the constraints

$$\rho \geq 0, \quad \int \rho \, d\Gamma = 1, \tag{2.7}$$

i.e., ρ is non-negative and normalized to 1. An alternative and equivalent point of view is to introduce the concept of an ensemble [Gibbs (1902)]. This is a collection of a large number of identical systems governed by the same Hamiltonian H; each system is at a different point in phase space, but otherwise is independent of the other systems. $\rho d\Gamma$ gives the number of systems in the volume element $d\Gamma$. The conservation of the probability or total number of systems of the ensemble can be expressed in the form of a differential equation as

$$\frac{\partial \rho}{\partial t} + \frac{\partial}{\partial \Gamma} \cdot \left(\dot{\Gamma} \rho \right) = 0. \tag{2.8}$$

Here, the distribution ρ is over the phase space point Γ and parametrically depends on time t. $\dot{\Gamma} = d\Gamma/dt$ is the "velocity" vector of the phase space point moving according to Hamilton's equations of motion, $\dot{\Gamma} = (\Gamma, H)$, and $\partial/\partial\Gamma = (\frac{\partial}{\partial q_1}, \frac{\partial}{\partial q_2}, \ldots, \frac{\partial}{\partial p_1}, \ldots)$ is the gradient operator in phase space. The second term in the equation is the divergence of the probability current in phase space. This equation is analogous to the continuity equation for charge conservation in electrodynamics, and mass conservation in fluid dynamics. To derive this equation, one can consider a small volume bounded by its surface and ask how systems move in and out of the volume if the systems are represented as points distributed in the phase space.

2.2.1   Liouville's equation

Equation (2.8) is a reflection of conservation of the total probability. However, the Hamiltonian system has a special property. Viewed as a collection of points in phase space for the probability distribution, these points behave like an incompressible fluid, satisfying

$$\frac{\partial}{\partial \Gamma} \cdot \dot{\Gamma} = 0. \tag{2.9}$$

In component form, the left-hand side of the equation is $\partial \dot{q}/\partial q + \partial \dot{p}/\partial p$. Substituting the Hamilton equations for the rate of change, we have $\partial^2 H/(\partial q \partial p) - \partial^2 H/(\partial p \partial q)$, which is obviously 0. Using this result, since $\frac{\partial}{\partial \Gamma} \cdot \left(\dot{\Gamma} \rho \right) = \rho \frac{\partial}{\partial \Gamma} \cdot \dot{\Gamma} + \dot{\Gamma} \cdot \frac{\partial \rho}{\partial \Gamma}$, the conservation of the probability becomes

$$\frac{D\rho}{Dt} \equiv \frac{\partial \rho}{\partial t} + \dot{\Gamma} \cdot \frac{\partial \rho}{\partial \Gamma} = 0. \tag{2.10}$$

The notation D is for the total derivative defined as the right-hand side of the equivalence sign $\equiv$. This equation is called the Liouville equation. We can also write, with the help of Hamilton's equations of motion and the Poisson bracket,

$$\frac{\partial \rho}{\partial t} = (H, \rho). \tag{2.11}$$

The total derivative $D\rho/Dt = 0$ means that the probability density does not change if we consider a small volume in phase space $d\Gamma$ such that the surface bounding the volume follows the motion of the systems. Since the density is the number of systems divided by the volume $d\Gamma$, and the points representing the systems cannot pass in or out of the surface because the surface is following the motion, we must have $d\Gamma$ not changing, which is the content of Liouville's theorem.

2.3   Microcanonical ensemble

Assuming we have a way of obtaining the probability distribution ρ, for example, by actually solving Liouville's equation, the average of a general quantity A can be computed as

$$\langle A \rangle = \int A\rho \, d\Gamma. \tag{2.12}$$

Alternatively, we can also solve the Hamilton equations of motion, obtaining the particle trajectories Γ_t. From this solution, we can also consider the time average

$$\bar{A} = \frac{1}{\tau} \int_0^\tau A(\Gamma_t) \, dt. \tag{2.13}$$

The latter approach is what we call "molecular dynamics," if we solve the equations numerically on a computer. In equilibrium statistical mechanics, we bypass the difficult time average and insist that the phase space average and time average are equal, i.e., we assume $\langle A \rangle = \bar{A}$ for sufficiently long time τ. However, this assertion in general cannot be proven rigorously, and we call it an ergodic hypothesis.

What form of the distribution should we take for the phase space average? Since we are only interested in equilibrium systems, we expect that ρ should be independent of time t. From Liouville's equation, this implies that $(\rho, H) = 0$ or ρ should be a function of the Hamiltonian H only. But if $H = U$ is the only conserved quantity with respect to the dynamics, we can see that $\rho = \text{const}$ is a very reasonable solution. There is one slight subtlety here. Since ρ needs to be integrated over the whole phase space volume, we cannot define ρ just on the energy surface $H(\Gamma) = U$ (we could however if we use the Dirac delta function). Also, the conservation of phase space volume, as per Liouville's theorem, requires a finite region of volume Ω. Thus, the micro-canonical ensemble [Huang (1987)] will be defined by

$$\rho(\Gamma) = \begin{cases} \text{const,} & \text{if } U \leq H(\Gamma) < U + \Delta, \\ 0, & \text{otherwise.} \end{cases} \tag{2.14}$$

See Fig. 2.1 for an illustration.

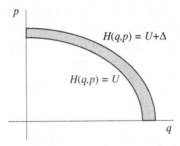

Fig. 2.1   For the microcanonical distribution in phase space, only the region defined by the thin shell bounded by energy U and $U + \Delta$ has nonzero probability and is uniform per phase space volume. Note that the thickness of the shell is not uniform and is proportional to $1/|\nabla H|$.

Using the normalization condition of ρ, the constant can easily be fixed to be the inverse of the phase space volume between energy U and $U + \Delta$. Here Δ is a small quantity in comparison with U. The assumption that all the microscopic states within a given energy interval have equal weight is also called the "principle of equal a priori probabilities."

2.3.1   *Boltzmann's principle*

Although we have a prescription to compute averages by Eq. (2.12), not all quantities can be computed this way. It is possible to express kinetic energy or pressure this way, but not entropy. A key equation to relate statistical mechanics to thermodynamics is Boltzmann's principle, which gives a formula to compute the thermodynamic entropy [Planck (1914)]:

$$S = k_B \ln \Omega, \quad \Omega(U, V, N) = \frac{1}{N! \, h^{3N}} \int_{U \leq H < U + \Delta} d\Gamma. \qquad (2.15)$$

The Boltzmann constant takes the value $k_B = 1.380\,649 \times 10^{-23}$ joule/kelvin. Here Ω is the number of states consistent with the given external constraints. We consider simple systems with the constraints of energy U, volume V, and number of particles N. How do we then count the number of states? In quantum mechanics, for a closed, finite system, energy is quantized, and the states are discrete. The counting of the states is very natural. In classical physics, we can still count states by dividing the phase space volume by h for each pair of (q, p), i.e., the phase volume should be counted in units of the action given by Planck's constant. To see this, we really need to treat the system quantum-mechanically, and take then the classical limit. The old quantum theory with the Bohr-Sommerfeld quantization, $\oint p\,dq = nh$, illustrates this phase space counting rule in a simple way. The factorial $N!$ is needed for a system of identical particles, but a justification (in order to avoid the Gibbs paradox) requires quantum mechanics. How do we know that the proposed formula for entropy is correct? In principle we can check it against Callen's postulates of entropy. Since we have proposed it as a principle, we will not do that, and focus on its consequences.

2.4 Sackur-Tetrode formula

As an application of the Boltzmann entropy formula, let us consider an ideal gas confined in a box of volume V. The Hamiltonian is a sum of free particle kinetic energies,

$$H = \sum_{j=1}^{3N} \frac{p_j^2}{2m},$$ (2.16)

provided that the coordinates q_j are within the volume. The first particle has the 3D vector momentum $\mathbf{p} = (p_1, p_2, p_3)$, the second particle (p_4, p_5, p_6), etc. Thus,

$$\Omega = \frac{1}{N!\,h^{3N}} \int_{U \leq H < U+\Delta} dq_1 dq_2 \cdots dq_{3N} dp_1 dp_2 \cdots dp_{3N}.$$ (2.17)

Here we have written out $d\Gamma$ with explicit coordinate and momentum variables. Since the Hamiltonian does not depend on q_j, the constraint is only on p_j. The q_j integrals can then be performed, producing a factor V^N. We get

$$\Omega = \frac{V^N}{N!\,h^{3N}} \left[\Upsilon(U + \Delta) - \Upsilon(U) \right].$$ (2.18)

We have defined the momentum integrals up to energy U as,

$$\Upsilon(U) = \int_{\sum_{j=1}^{3N} \frac{p_j^2}{2m} \leq U} dp_1 dp_2 \cdots dp_{3N} = (2mU)^{3N/2} \frac{\pi^{3N/2}}{(3N/2)!}.$$ (2.19)

The value of this multi-dimensional integral can be obtained by a change of variable from p_j to $x_j = p_j/\sqrt{2mU}$. In doing this, we can transform the problem into finding the volume of a unit hypersphere in n dimensions. The result can be demonstrated to be (see the math reading Sec. 2.7)

$$c_n = \int_{x_1^2 + x_2^2 + \cdots + x_n^2 \leq 1} dx_1 dx_2 \cdots dx_n = \frac{\pi^{n/2}}{(n/2)!}. \tag{2.20}$$

Here the factorial function is defined through the Gamma function [Arfken (1970)] $\Gamma(x+1) = x!$. This gives $c_1 = 2$, $c_2 = \pi$, and $c_3 = 4\pi/3$ for a length segment, circle, and sphere of unit radius, as expected.

2.4.1 *Thermodynamic limit*

With the Boltzmann principle, Eq. (2.15), and the explicit expression for the phase space volume, Eq. (2.18) and (2.19), we have in principle obtained the formula for entropy. However, we are still one step away from deriving the thermodynamic entropy. The reason is that the current formula as it is does not satisfy the extensivity condition, $S(\lambda U, \lambda V, \lambda N) = \lambda S(U, V, N)$ as implied by Callen's postulates. This property is achieved by taking the so-called thermodynamic limit. By this, we mean that N is large, and similarly, V and U are large, such that the ratio U/N or V/N is fixed. As a result, the entropy S will be proportional to N. Not all statistical-mechanical systems have such a property, for example, systems with long-range gravitational interactions. When the thermodynamic limit does exist, we call such a system 'normal' [Kubo (1965)].

In the following derivation, we will use the Stirling approximation,

$$\ln N! = N \ln N - N + O(\ln N), \tag{2.21}$$

and drop any terms that are small in comparison to N, such as terms of order 1, or even $\ln N$. We also assume that Δ is small, so that we can use $\Upsilon(U + \Delta) - \Upsilon(U) \approx \Upsilon'(U)\Delta$. With this approximation made, we find

$$S = k_B \ln \left[\left(\frac{V(2\pi mU)^{3/2}}{h^3} \right)^N \frac{\Delta}{U} \frac{1}{N! \, (3N/2 - 1)!} \right]. \tag{2.22}$$

Using the property of the logarithm function, $\ln(ab) = \ln a + \ln b$, we find that the first factor is proportional to N; the second factor $\ln(\Delta/U)$ will be dropped, as it is indeed small if Δ is kept finite or order 1. For the last two factorial terms, we can simplify using the Stirling formula. After regrouping and keeping only terms of order N or $N \ln N$, dropping terms of $\ln N$ or smaller, we finally obtain

$$S = k_B N \ln \left[\frac{V}{N^{5/2}} \left(\frac{4\pi m U}{3h^2} \right)^{3/2} \right] + \frac{5}{2} k_B N. \tag{2.23}$$

This is the entropy for the ideal gas, derived circa 1912 independently by Otto Sackur and Hugo Tetrode [Grimus (2011)]. Unlike the thermodynamic method, which is determined by the Clausius equation ($\int \delta Q / T$) only up to an arbitrary additive constant, here the entropy is determined absolutely, and it involves a quantum feature, the Planck constant h.

2.4.2 *Equations of states*

From the thermodynamic relation, $dU = TdS - pdV + \mu dN$, or more conveniently for the entropy, $dS = dU/T + pdV/T - \mu dN/T$, we find

$$\frac{\partial S}{\partial U} = \frac{1}{T} = \frac{\partial}{\partial U} \left(k_B N \ln U^{3/2} + \text{const} \right) = \frac{3k_B N}{2U}. \tag{2.24}$$

Or $U = \frac{3}{2} N k_B T$, which gives the heat capacity at constant volume to be $C_V = dU/dT = \frac{3}{2} N k_B$. The factor 3/2 is because we assume that our molecules are mono-atomic. Similarly,

$$\frac{\partial S}{\partial V} = \frac{p}{T} = \frac{N k_B}{V}, \quad \rightarrow \quad pV = N k_B T, \tag{2.25}$$

which is the ideal gas law. Finally, the chemical potential of the ideal gas can be derived from the derivative of the entropy with respect to the particle number N, obtaining

$$\mu = k_B T \ln \left[\frac{N \lambda^3}{V} \right], \quad \lambda = \frac{h}{\sqrt{2\pi m k_B T}}. \tag{2.26}$$

Here we have introduced a quantity λ which is known as the thermal wavelength.

As claimed, the entropy function $S(U, V, N)$ contains all the thermodynamic information we need to know. This single function includes the information about the equation of states, heat capacity, etc. Thus, the mission of equilibrium statistical mechanics is accomplished if this function is known. But now a question. Does the entropy expression, Eq. (2.23), we computed satisfy all of Callen's postulates? Unfortunately, the answer is no. We should ponder why?

2.5 From microcanonical ensemble to canonical ensemble

In this section, we give a derivation of another perhaps more convenient ensemble for practical calculations — the canonical ensemble. The basic idea is to consider two systems coupled in such a way so that energies can be distributed between the two. As a result, the energy of a subsystem is no longer a fixed value (or in a narrow window of $[U, U + \Delta)$), but can fluctuate. In this setup, particles among the two are not allowed to redistribute. However, one of the systems, say, system 2, will be taken to be very large and our focus will be on system 1. System 2 will be called the thermal bath.

In this derivation, for the purpose of representing the phase distribution ρ of a microcanonical ensemble, it is mathematically convenient to take the limit of Δ, the energy window, to 0. Since we still need the normalization, $\int \rho \, d\Gamma = 1$, the value of ρ on the energy surface $H(\Gamma) = U$ must be very large. We can represent this situation with the Dirac-δ function,

$$\rho(\Gamma) = C\delta\big(H(\Gamma) - U\big), \quad \frac{1}{C} = \int \delta\big(H(\Gamma) - U\big)d\Gamma. \qquad (2.27)$$

The constant C is fixed by normalization. Due to the δ-function constraint, we could write the phase space average as a surface integral over a $(3N - 1)$ dimensional surface defined by $H(\Gamma) = U$. However, the distribution on the surface is, in general, not uniform, but is proportional to $d\sigma/|\partial H/\partial \Gamma|$ [Khinchin (1949)], here $d\sigma$ is the surface area element, and we need to divide it by the magnitude of the gradient vector of the Hamiltonian in phase space.

Let us split the Hamiltonian into $H(\Gamma) = H_1(\Gamma_1) + H_2(\Gamma_2) + V_{12}$, where the overall variables $\Gamma = (\Gamma_1, \Gamma_2)$ are split into two groups: system 1 depending only on Γ_1, and system 2 on Γ_2. In order for the energy to be transferrable between the two, we also need an interaction term V_{12}. However, after its effect is done, we will turn this infinitesimally small. In practice, in the following derivation we set $V_{12} = 0$.

System 2 will be considered an environment, and its details are of no interest to us. We would like to characterize system 1 without knowing the microscopic states of system 2. Thus, the probability distribution of system 1 is given by the marginal distribution of the overall joint distribution $\rho(\Gamma)$,

$$\rho_1(\Gamma_1) = \int \rho(\Gamma_1, \Gamma_2)d\Gamma_2 = C \int \delta\big(H(\Gamma_1, \Gamma_2) - U\big)d\Gamma_2. \qquad (2.28)$$

Here the integration is over system 2 only, fixing system 1 to a particular state Γ_1. Using the energy decomposition, $H = H_1 + H_2$, we can write the integral of the delta function as

$$\int \delta\big(H(\Gamma_1, \Gamma_2) - U\big) d\Gamma_2 = \int \delta\big(H_2(\Gamma_2) - U + H_1\big) d\Gamma_2$$

$$= \frac{1}{C_2(U - H_1)}, \qquad (2.29)$$

which is nothing but the inverse of the normalization constant for system 2 at fixed energy $U - H_1(\Gamma_1)$ (remember that we hold Γ_1 constant). The normalization constants for the microcanonical ensembles are closely related to the phase space volume defined as Ω. In fact, we can write $\Omega(U) \approx \Delta/(CN!h^{3N})$. Here we take back a finite Δ so that Ω has the correct dimension. We thus obtain an important result

$$\rho_1(\Gamma_1) \propto \Omega_2\big(U - H_1(\Gamma_1)\big), \qquad (2.30)$$

i.e., the distribution of system 1 is proportional to the number of microscopic states at energy $U - H_1$ in system 2. The proportionality constant in the above equation can be fixed by the probability normalization, $\int \rho_1 d\Gamma_1 = 1$.

Without making further assumptions, we may be stuck. The next step is to assume that the second system is large in comparison with the first one. We can then make a Taylor expansion in the small variable $H_1 - \langle H_1 \rangle$. However, Ω_2 is a very fast varying function of its argument. The correct expansion is not Ω_2 but its logarithm, i.e.,

$$\ln \Omega_2(U - H_1) = \ln \Omega_2(\langle H_2 \rangle) - \frac{\partial \ln \Omega_2(U)}{\partial U}\bigg|_{U = \langle H_2 \rangle} \big(H_1 - \langle H_1 \rangle\big)$$

$$+ \frac{1}{2} \frac{\partial^2 \ln \Omega_2(U)}{\partial U^2}\bigg|_{U = \langle H_2 \rangle} \big(H_1 - \langle H_1 \rangle\big)^2 + \cdots. \qquad (2.31)$$

Here the total energy $U = \langle H_1 \rangle + \langle H_2 \rangle$ is conserved. Since $S_2 = k_B \ln \Omega_2$ is the entropy of system 2, the coefficient of the second term is precisely $1/(k_B T_2)$, and the next order term is proportional to the inverse of the heat capacity of the second system (show this). Since we assume the second system is very large compared to the first system, we can drop the high order terms. Finally, we can write

$$\rho_1(\Gamma_1) \propto e^{-\beta H_1(\Gamma_1)} \qquad (2.32)$$

Here we have defined $\beta = 1/(k_B T_2)$. Notice that the temperature refers to that of the system 2, but in the following, we will just speak of the temperature of 'the' system, which is system 1. The proportionality constant can be fixed by $\int \rho_1 d\Gamma_1 = 1$.

2.5.1 *Connection to thermodynamics*

Having derived the canonical distribution, $\rho_1 \propto \exp(-\beta H_1)$, also known as the Gibbs distribution [Gibbs (1902)], we shall omit the index 1 for simplicity. Just as with the microcanonical distribution, we can make a connection to thermodynamics. In the microcanonical ensemble, that connection is the phase space volume counted in units of h^{3N} relating to the thermodynamic entropy through the Boltzmann principle. We would like to show here that in the canonical ensemble, the relevant thermodynamic function is not entropy, but the Helmholtz free energy $F = U - TS$. To establish such a connection, first we define the partition function, as

$$Z = \int \frac{d\Gamma}{N!h^{3N}}\, e^{-\beta H(\Gamma)}. \tag{2.33}$$

Here we assume that each particle is moving in three-dimensional space, and the $N!$ term is needed to take care of the indistinguishable nature of the particles. The integration is carried out in the whole phase space without any constraints, and $d\Gamma = dq_1 dq_2 \cdots dq_{3N} dp_1 dp_2 \cdots dp_{3N}$.

Since we already know that phase space volume is related to the entropy, we would like to express Z in terms of what is known. To this end, we insert $1 = \int \delta(E - H(\Gamma)) dE$ into the integral, and then exchange the order of integration, we get

$$\begin{aligned} Z &= \int \frac{d\Gamma}{N!h^{3N}} e^{-\beta H(\Gamma)} \\ &= \int dE\, e^{-\beta E} \int \frac{d\Gamma}{N!h^{3N}} \delta\big(E - H(\Gamma)\big) \\ &= \int dE\, e^{-\beta E + S(E)/k_B} \frac{1}{\Delta}. \end{aligned} \tag{2.34}$$

Here we have used the fact that $\Omega(E) \approx \omega(E)\Delta$, and $\omega(E) = \int \frac{d\Gamma}{N!h^{3N}} \delta\big(E - H(\Gamma)\big)$ is the density of states, and re-expressed it using entropy by the Boltzmann principle, $e^{S/k_B} = \omega(E)\Delta$. The entropy is also a function of other variables, such as V and N. Since they are fixed with respect to the integration, we have suppressed them. In order to be able to evaluate the E integration, we invoke the fact that the number of particles N is large.

In the large N limit, the first factor $e^{-\beta E}$ decays rapidly with E while the second factor of density of states $\omega(E)$ grows rapidly with E. As a result, the integrand is sharply peaked around the average value of E. To show this quantitatively, we introduce the energy per particle $\epsilon = E/N$, and use the fact that $S = Ns(\epsilon)$ is extensive, where s is the entropy per particle. Then

$$Z = \frac{N}{\Delta} \int d\epsilon\, e^{-N\beta(\epsilon - Ts)}. \tag{2.35}$$

Asymptotic results of such an integral for large N can be obtained if we Taylor expand the factor on the exponential around its minimum, $f(\epsilon) = \epsilon - Ts = \epsilon_0 - Ts(\epsilon_0) + \frac{1}{2}f''(\epsilon_0)(\epsilon - \epsilon_0)^2 + \cdots$. This is well justified in the large N limit, as most of the contributions are around the minimal value of f. Putting this expansion back into the integral, we see that the integrand is a Gaussian function peaked around ϵ_0 with a width proportional to $1/\sqrt{N}$. It is obvious that the minimal value is also the average value, $E_{\min} = \langle H \rangle = \epsilon_0 N$. Performing the Gaussian integration, we find

$$Z = \frac{N}{\Delta} e^{-\beta N\left(\epsilon_0 - Ts(\epsilon_0)\right)} \sqrt{\frac{2\pi}{\beta N f''(\epsilon_0)}}. \tag{2.36}$$

The minimum of energy ϵ is determined by the first derivative, $f'(\epsilon_0) = 0$, or $1 - Tds/d\epsilon = 0$, or equivalently, at $E_{\min} = N\epsilon_0$ determined by the equation $1/T = \partial S/\partial U$. Thus, we find

$$-\frac{1}{\beta} \ln Z = \min_E \left(E - TS(E, V, N)\right) + O(\ln N)$$

$$= F(T, V, N). \tag{2.37}$$

That is, $-k_B T \ln Z$ can be identified with the thermodynamic function of Helmholtz free energy as a Legendre transform of the entropy function. The entropy is a function of U, V, and N, but the free energy is a function of T, V, and N. To make a change of variable from U to T, one minimizes the expression $U - TS(U, V, N)$ with respect to U fixing T, V, and N. This is equivalent to inverting the equation $1/T = \partial S/\partial U$ to view U as a function of T, V, and N instead.

2.5.2 *Harmonic oscillators*

The above derivation clearly relies on the fact that N is large. The equivalence of the microcanonical ensemble and canonical ensemble is true only in the thermodynamic limit. In this subsection, we illustrate this fact with a simple example of N independent classical oscillators, with the Hamiltonian

$$H = \sum_{j}^{N} \left(\frac{p_j^2}{2m} + \frac{1}{2} m\omega_0^2 q_j^2 \right), \tag{2.38}$$

where all oscillators have the same mass m and intrinsic angular frequency ω_0. A quantum version of this model was proposed by Einstein as a model for solids to explain the decreasing heat capacity as one approaches low temperatures. The classical model recovers what is known as the Dulong-Petit law (each atom in a solid contributes $3k_B$ to heat capacity). In the canonical ensemble, since each oscillator is independent of the others, we can write, $Z = z^N$, with

$$z = \int \frac{dq\,dp}{h} e^{-\beta\left(\frac{p^2}{2m} + \frac{1}{2}m\omega_0^2 q^2\right)} = \frac{1}{\beta\hbar\omega_0}. \tag{2.39}$$

The free energy is given by $F = -k_B T \ln Z = \frac{N}{\beta} \ln(\beta\hbar\omega_0)$. There is no $N!$ factor in Z, why?

From the free energy, we can calculate entropy as $S = -\partial F/\partial T = k_B N(1 + \ln \frac{k_B T}{\hbar\omega_0})$, internal energy $U = F + TS = Nk_B T$, and the heat capacity $C = dU/dT = Nk_B$.

Using the microcanonical ensemble, we need to compute the phase space volume. Let us define

$$\Gamma(U) = \frac{1}{h^N} \int_{H \leq U} dq_1 dq_2 \cdots dq_N dp_1 dp_2 \cdots dp_N. \tag{2.40}$$

This integral can be evaluated if we scale the coordinates and momenta so that the constraint equation is a hypersphere. As its volume can be found from the formula c_n with $n = 2N$, we get

$$\Gamma(U) = \frac{1}{N!} \left(\frac{U}{\hbar\omega_0} \right)^N. \tag{2.41}$$

Here the factorial $N!$ has nothing to do with identical particles, rather, it is from the volume of the hypersphere. According to Boltzmann's principle, we need to set, $S = k_B \ln\big(\Gamma(U + \Delta) - \Gamma(U)\big)$. However, we will use $S = k_B \ln \Gamma(U)$ (known as Gibbs volume entropy), and they differ by a small amount of $O(\ln N)$ only. In any event, if we do not use the Stirling approximation, which throws away terms of order $\ln N$, we see that the micro-canonical entropy and canonical entropy do not agree. If we assume that these differences are unimportant for large systems, then we find

$$S = k_B N \ln \left(\frac{U}{\hbar \omega_0 N} \right) + k_B N, \qquad (2.42)$$

which is identical to the canonical result, since $1/T = \partial S/\partial U = k_B N/U$, or $U = k_B T N$.

2.6 Grand canonical ensemble

In a grand canonical ensemble, we allow for the particle numbers to fluctuate in addition to energy. Thus, the thermal bath attached to the system exchanges particles as well as energy. We will forgo the details and only present the key equations. Interested readers can refer to Huang [Huang (1987)]. The distribution function in the grand canonical ensemble will depend on the particle number N which is a variable, as well as the usual phase space point Γ for each given N. It is

$$\rho \propto e^{-\beta \left(H_N(\Gamma) - \mu N \right)}. \qquad (2.43)$$

Here H_N is the Hamiltonian when the system has N particles. We define the grand partition function as

$$\Xi = \sum_{N=0,1,2,\cdots} \int \frac{d\Gamma}{N! h^{3N}} e^{-\beta(H_N - \mu N)} = \sum_{N=0}^{\infty} Z_N e^{\beta \mu N}. \qquad (2.44)$$

Here Z_N is the canonical partition function of N particles. Note that we start the sum with 0 particles, with the definition $Z_0 = 1$. The connection with thermodynamics is through the grand potential,

$$\Psi = U - TS - \mu N = -pV = -k_B T \ln \Xi(T, V, \mu). \qquad (2.45)$$

In thermodynamics, this is again a Legendre transform from the particle number N to the chemical potential μ. We thus have the important thermodynamic differential relation

$$d\Psi = -SdT - pdV - \langle N \rangle d\mu. \qquad (2.46)$$

To avoid confusion with the variable N, we write the thermodynamic particle number as an ensemble average of the grand canonical ensemble. The grand canonical ensemble is particularly convenient when we study phase transitions or when we are dealing with fermions in quantum statistical mechanics.

2.7   Mathematical preliminaries — Gamma function, Gaussian integral, and volume of a hypersphere

The Gamma function is defined as

$$\Gamma(z+1) = z! = \int_0^\infty e^{-t} t^z dt. \tag{2.47}$$

It satisfies the recursion relation $\Gamma(z+1) = z\Gamma(z)$. Repeated application of this relation shows that it is indeed the usual factorial function for integers $z \geq 0$. But the integral representation is valid for any complex number with $\mathrm{Re}(z+1) > 0$. With the relation $\Gamma(z)\Gamma(1-z) = \pi/\sin(\pi z)$, the Gamma function is in fact defined for the whole complex plane with simple poles at zero and negative integer values. Using the steepest descent method to perform the integral, we can obtain an asymptotic expansion result for large z [Arfken (1970)]:

$$\ln n! = n\ln n - n + \frac{1}{2}\ln n + \ln\sqrt{2\pi} + \frac{1}{12n} - \frac{1}{360n^3} + \cdots. \tag{2.48}$$

Here we give a 'quick and dirty' derivation of the Stirling formula which is correct only for the first two terms. If we take the logarithm of the factorial function, we obtain a sum of $\ln i$ which can be converted to an approximate integral for large i.

$$\ln n! = \sum_{i=1}^n \ln i \approx \int_1^n \ln x \, dx = x\ln x \Big|_1^n - \int_1^n x\, d(\ln x) \approx n\ln n - n. \tag{2.49}$$

Here we have used integration by parts for the integral. If we do the resulting integral, we obtain $n-1$. The -1 term is omitted, as it is incorrect when compared to the more accurate result.

The Gaussian integral, defined as

$$I = \int_{-\infty}^{+\infty} e^{-\frac{x^2}{2}} dx, \tag{2.50}$$

can be performed if we take two copies of it and multiply them together, and call the integration variables x and y. The two-dimensional integral can be transformed to polar coordinates, $x = r\cos\theta$, $y = r\sin\theta$, $dx\,dy = r dr\,d\theta$, and the resulting integral can be carried out:

$$I^2 = \int_{-\infty}^{+\infty} e^{-\frac{x^2}{2}} dx \int_{-\infty}^{+\infty} e^{-\frac{y^2}{2}} dy = \int_0^\infty e^{-r^2/2} r dr \int_0^{2\pi} d\theta = 2\pi. \tag{2.51}$$

Thus, $I = \sqrt{2\pi}$.

Finally, the volume of a hyper-sphere can be calculated with the help of Gaussian integrals and the Gamma function. Consider the volume of a hyper-sphere of radius R in n dimensions.

$$\Omega_n = \int_{\sum_{i=1}^{n} x_i^2 < R^2} dx_1 dx_2 \cdots dx_n = c_n R^n. \qquad (2.52)$$

The surface area of the sphere can be obtained by differentiating with respect to R, $S_n = d\Omega_n/dR = nc_n R^{n-1}$, since the surface area times a thickness of a shell, $S_n dR$, gives the volume of the shell. Now we take n copies of the Gaussian integral (without the factor of $1/2$), we find

$$\pi^{n/2} = \int_{-\infty}^{+\infty} dx_1 \cdots \int_{-\infty}^{+\infty} dx_n \, e^{-x_1^2 - x_2^2 - \cdots - x_n^2} = \int_0^{\infty} dR \, S_n(R) e^{-R^2}.$$
$$(2.53)$$

Here we have made an n-dimensional spherical polar transform. Since the integrand is symmetric, we just obtain the surface area times dR. The integrand can be put in the form of the Gamma function. With some algebra, we obtain

$$c_n = \frac{\pi^{n/2}}{(n/2)!}. \qquad (2.54)$$

We have $(1/2)! = \sqrt{\pi}/2$, this give $c_1 = 2$, $c_2 = \pi$, $c_3 = 4\pi/3$, as expected.

2.8 Supplementary reading — volume vs. area measure in the microcanonical ensemble

For the microcanonical distribution, if we take the thickness of the energy shell in phase space, Δ, to 0, it seems reasonable to say that the system can only stay on the surface. We might suggest to use the logarithm of the area of a hypersurface of constant energy in phase space as an expression for the thermodynamic entropy [Franzosi (2018)]. I show here that there are a number of difficulties for such an assignment. The area measure is not a Liouville invariant, and is not an invariant with respect to canonical transforms. Also, equipartition theorem does not hold, and the measure is not stationary with respect to the dynamics.

Let's take a one-degree harmonic oscillator to show some of the possible problems with the alternative definition of entropy S. The model is

$$H = \frac{P^2}{2M} + \frac{1}{2}M\Omega^2 x^2. \qquad (2.55)$$

Let U be the conserved energy, $U = H(P, x)$. Then the "surface" is a line in the phase space (x, P). I denote the length element by dl, and the usual microcanonical measure is $dl/|\nabla H|$ here

$$|\nabla H| = \sqrt{\dot{x}^2 + \dot{P}^2} = \sqrt{(P/M)^2 + (M\Omega^2 x)^2}. \qquad (2.56)$$

The first comment is that $\Delta dl/|\nabla H|$ is Liouville invariant by Liouville's theorem [Khinchin (1949)], and the length (in general $N-1$ dimensional surface in an N dimensional phase space) is not an invariant. Here Δ is the shell thickness defining the two surfaces, $H = U$ and $H = U + \Delta$.

The length measure does not have correct physical units. Since $dl = \sqrt{dx^2 + dP^2}$, the dimensions are not consistent and result depends explicitly on the choice of units. However, the usual $dl/|\nabla H|$ has no problem, as the inconsistency is canceled in between the numerator and denominator. This means if we use dl to define entropy, the entropy does not have meaningful units.

We can parameterize the curve by the solution to the problem as

$$x = \sqrt{2U/(M\Omega^2)} \cos(\Omega t), \qquad (2.57)$$

$$P = M\dot{x} = -\sqrt{2UM} \sin(\Omega t). \qquad (2.58)$$

The full ellipse is covered when t varies from 0 to $2\pi/\Omega$. Then, we see after some algebra that $dl/|\nabla H| = dt$, but

$$dl = \sqrt{\left(\cos^2(\Omega t) + \sin^2(\Omega t)/(M^2\Omega^2)\right)2MU\Omega^2}\, dt. \qquad (2.59)$$

Equipartition does not hold. One can check that, if we use the measure $dl/|\nabla H|$, equal partition does hold in the sense

$$\langle P^2/(2M) \rangle = U/2, \qquad (2.60)$$

$$\langle M\Omega^2 x^2/2 \rangle = U/2, \qquad (2.61)$$

$$\text{here } \langle \cdots \rangle = \frac{\int dl \cdots /|\nabla H|}{\int dl/|\nabla H|}. \qquad (2.62)$$

Note that since $dl/|\nabla H| = dt$, the average in phase space is the same as the average over time — ergodicity holds explicitly.

Equipartition does not hold with the measure dl without dividing by $|\nabla H|$. I can do an explicit calculation with $\Omega = U = 1$, but $M = 2$, I find numerically,

$$\langle P^2/(2M) \rangle_l = 0.42007, \tag{2.63}$$

$$\langle M\Omega^2 x^2/2 \rangle_l = 0.57993, \tag{2.64}$$

$$\text{now} \quad \langle \cdots \rangle_l = \frac{\int dl \cdots}{\int dl}. \tag{2.65}$$

The sum is 1, of course, since it must be equal to U by Eq. (2.55).

The surface measure without dividing the magnitude of the gradient of the Hamiltonian is not stationary. The standard microcanonical ensemble can be represented by the Dirac delta function (in the limit as the shell thickness Δ goes to 0) as $\rho_{\text{micro}} = C\delta(H - U)$. The probability distribution with the surface measure is then

$$\rho(P, x) = C|\nabla H|\delta(H - U). \tag{2.66}$$

C is fixed by normalization. The extra $|\nabla H|$ here is used to cancel the usual one in the denominator of the phase space surface element. Clearly, this distribution is in general not stationary:

$$\frac{\partial \rho}{\partial t} = (H, \rho) \neq 0. \tag{2.67}$$

Therefore, it cannot be used for equilibrium statistical mechanics. The parentheses denote the Poisson bracket. Since $|\nabla H|$ does not commute with H, as can be seen from the explicit equation, Eq. (2.56), and H is the only conserved quantity for the harmonic oscillator, we have $\partial\rho/\partial t \neq 0$.

Lastly, the surface measure is not a canonical transform invariant. The volume measure (the phase space volume) is invariant with respective to a canonical transform. But the surface area is not. For the 1D harmonic oscillator, we can do $\tilde{X} = \sqrt{M}x$, and then $\tilde{P} = P/\sqrt{M}$, which is a canonical transform, but the line element obviously changes. Alternatively, we can use action-angle variables (J, θ), $J = \oint Pdx/(2\pi)$, and then $H = \Omega J$. We can see that under this transform, the usual $dl/|\nabla H|$ is still $dt \propto d\theta$, but the line element dl changes.

2.9 Supplementary reading — origin of Boltzmann's principle

Boltzmann's 1877 paper [Sharp and Matschinsky (2015)] now has an English translation (Entropy, **17**, 1971–2009 (2015)). We give some historical background of Boltzmann's approach to entropy. His motivation was to give a microscopic and mechanical interpretation of thermodynamic entropy, but his system was a collection of classical identical particles with little or no interactions, i.e., ideal gas. First, essentially, Boltzmann has his principle in this paper:

$$S = k_B \ln W, \tag{2.68}$$

where W, a positive integer, is the number of microscopic states given the constraints. One of the constraints must be the energy U. In this sense, since U is fixed, the entropy above is defined for the microcanonical ensemble. Anticipating quantum mechanics, this formula has the advantage that a pure quantum state in the ground state has $S = 0$, fixing the absolute scale of entropy. However, this is not what Boltzmann originally applied this formula for. He considered further constraints, $\{n_i\}$, the number of molecules having a specific property, namely, the single particle energy (or equivalently position $\mathbf{r}$ and momentum $\mathbf{p}$ of a single particle), with a discretized description and later he made it continuous. He then wrote down the permutation formula:

$$W = \frac{N!}{\prod_i n_i!}. \tag{2.69}$$

Each single particle state has a specified occupation number n_i, and the formula gives the count of the number of possible arrangements, assuming the particles have labels. I say that S defined by W is for constrained equilibrium states, in the sense of Callen or Lieb, as equal probability for all the microstates is implied in the counting. Each state is given an equal weight as long as it is consistent with the constraints.

If we normalize by the total possible occupation numbers consistent with the additional constraints, i.e. $\sum n_i = N$, and $\sum \epsilon_i n_i = U$ where N is the total number of particles and U is the total energy, then we can say that S is the logarithm of probability of the distribution in equilibrium. We say in equilibrium of the microcanonical ensemble as each state is counted with equal weight. That is, the equilibrium distribution

$$P_{\text{eq}}(\{n_i\}) = \frac{W(\{n_i\})}{\sum_{\{n_j\}} W(\{n_j\})}. \tag{2.70}$$

For the rest of the derivation Boltzmann was trying to change from the microcanonical ensemble to a canonical ensemble of a single particle. This gives the most probable value of n_i as the usual Boltzmann distribution $n_i \propto \exp(-\beta\epsilon_i)$, by maximizing W or $\ln W$, subject to the overall constraints, while adjusting the occupation at state i.

Using the Stirling approximation, we can also write S as

$$S = -k_B \int d\mathbf{r} d\mathbf{p} \, f \ln f + \text{const.} \tag{2.71}$$

where $n_i = f d\mathbf{r} d\mathbf{p}$, making a transition from discrete to continuous description.

As we know, f satisfies the celebrated Boltzmann equation he derived in 1872. With the H-theorem, entropy increase is proved. So in this formula, entropy indeed can be generalized to time-dependent cases, $S(t)$, since f depends on time.

Swendsen's definition [Swendsen (2012)] almost parallels Boltzmann's except that Swendsen works at an even higher level of description, i.e., the constraints are not the occupation numbers but the energies E_j and additional conserved quantities of subsystems of a composite system. This parallels Callen's formulation of thermodynamics very well.

I must emphasize that Swendsen's definition of entropy is also based on the equilibrium distribution of probability of energies, $P_{\text{eq}}(\{E_j\})$, and not a nonequilibrium evolution of the probability of the actual system. So then, what is the meaning of $k_B \ln P_{\text{eq}}(\{E_j\})$? I think it should be interpreted as the total entropy (up to a constant we don't care) of the constrained equilibrium system, when the constraints are enforced. The maximum value is the value if the constraints are removed, consistent with Callen's postulates of entropy.

Then how do we evolve the system dynamically? One may easily fall into the trap that the change of the entropy in time is

$$S(t) = \max_{E_j} k_B \ln P(\{E_j\}, t), \tag{2.72}$$

i.e., look at the peak value of the actual time dependent probability distributions, and take the log. Clearly this gives nonsensical results. Because it usually decreases, not increases. However, $S(t \to \infty)$ gives correct results (apart from how to fix the normalization). So the way close to Boltzmann's approach is to use

$$S(t) = k_B \ln P_{\text{eq}}(\{E_j(t)\}), \tag{2.73}$$

where the distribution function of equilibrium is used but the values of energies vary with time. We can use the average values of energies in the nonequilibrium ensemble $P(\{E_j\}, t)$ or the most probable value of energies at time t. With this interpretation, we get the correct behavior that entropy increases with time.

The Boltzmann principle, formula (2.68), is also used with the least possible constraints, i.e., with only the total energy U (or in a window U and $U + \Delta$) fixed. When used in this way, clearly, it is a quantity in equilibrium only. With this understanding and knowing that it cannot be applied to nonequilibrium systems since nothing can be changed (unless one subdivides the system for a slightly more detailed description of the system), it is perfectly fine as a definition for equilibrium entropy. This is the point of view we use in this textbook.

Formula (2.68) also has the nice feature when applied to constrained composite systems, where the counts are multiplicative, and entropies are thus additive. Swendsen's definition of entropy is strictly additive when the systems are relaxed at equilibrium values of the unconstrained energies, but formula (2.68) will not be, unless sizes are large (this is a correction of order $1/N$, i.e., log sum of all states with fixed E and that from the most probable $\{E_j\}$ differ by an amount of order less than N).

For the nonequilibrium situation, we can consider again the composite systems and then determine $E_j(t)$ as function of time t. We can determine the entropy of the composite systems based on formula (2.68), the Boltzmann principle, for the constrained equilibrium problem, and then plug in the actual observed value of E_j back into it.

The translators' commentary notes that Boltzmann's approach does not have the Gibbs $N!$ problem. That is not true, if he is careful not to drop the $N \ln N$ constant from the logarithm of the number of permutations to the permutability measure. The correct counts that are consistent with extensitivity need to remove the $N!$ to have

$$W = \prod_i \frac{g_i^{n_i}}{n_i!}, \tag{2.74}$$

where g_i is the number of allowed states in cell i, proportional to the single particle phase space volume of the cell, in a classical context. Dividing out the $N!$ is known as the 'Boltzmann counting'. We recover the Sackur-Tetrode formula if we take $g_i \equiv dxdydzdp_xdp_ydp_z/h^3$ with h the Planck constant.

Problems

Problem 2.1. *Consider the classical ideal gas system of N point particles with the Hamiltonian*

$$H = \sum_{j=1}^{3N} \frac{p_j^2}{2m} + V(q_1, q_2, \ldots, q_{3N}).$$

The effect of the potential energy V is to confine the particles to a box of volume $V = L^3$, and to establish equilibrium. It is treated as 0 otherwise. Study the ideal gas in three different ensembles: micro-canonical, canonical, and grand-canonical. That is, compute the micro-canonical number of states $\Omega(U,V,N)$, the canonical partition function $Z(T,V,N)$, and grand-canonical partition function $\Xi(T,V,\mu)$. Differentiate with respect to appropriate arguments of the fundamental thermodynamic relations [micro-canonical entropy $S(U,V,N)$, the Helmholtz free energy $F(T,V,N)$ in canonical ensemble, and grand potential $\Psi(T,V,\mu)$] to get the equations of states: (1) internal energy U as a function of temperature T, (2) the ideal gas law $pV = Nk_BT$, and (3) the chemical potential μ, or (4) average number of particle $\langle N \rangle$ in the grand-canonical ensemble. Show that the three ensembles give consistent (identical) results for all the quantities calculated in the thermodynamic limit.

Problem 2.2. *Consider N non-interacting point particles moving between a fixed, permeable, and diathermal wall separating two compartments of volume V_1 and V_2 with $V_1 = V_2 = V$. The Hamiltonian of the combined system can be written as*

$$H = \sum_{i=1}^{N} \left(\frac{\mathbf{p}_i^2}{2m} + u(\mathbf{r}_i) \right).$$

The single particle potential energy $u(\mathbf{r})$ is $+\infty$ for $\mathbf{r}$ outside the two compartments, and is 0 in the first compartment, but is a constant u_0 in the second compartment. (a) Are the temperatures of the gas in the two compartments equal? (b) Which of the ensembles among the microcanonical, canonical, and grand-canonical is most suited for this problem? (c) Calculate the average number of particles N_1 and N_2 in each compartment, expressed in terms of the temperature T of the compartment, the potential u_0, and the total number of particles N. (d) Compute the pressure p_1 and p_2 in each compartment.

Problem 2.3. *Heat capacities can behave differently in the canonical ensemble and micro-canonical ensemble in a finite system of N degrees of freedom. We elaborate this point with the following questions. (a) Express the heat capacity, $C_1 = dU/dT$, in terms of the fluctuation of the energy, $\langle H^2 \rangle - \langle H \rangle^2$, where the average is over the canonical distribution. Other model parameters, such as system volume, external field, etc., are fixed. Show that $C_1 \geq 0$. (b) Suppose that the entropy is calculated as a function* *of energy as, $S = S(U)$, in a micro-canonical ensemble. Derive a formula relating the heat capacity C_2 to the entropy function. (c) Discuss the condition for $C_1 = C_2$. Is it possible that $C_2 < 0$, and why?*

Problem 2.4. *Consider a one-dimensional classical harmonic oscillator with the Hamiltonian*

$$H = \frac{p^2}{2m} + \frac{1}{2}m\omega^2 q^2.$$

(a) Compute the partition function Z of the canonical ensemble, the Helmholtz free energy F, the entropy S, and the internal energy U as functions of temperature T. (b) Compute for the micro-canonical ensemble the number of microstates using two different definitions for the number of microstates and entropy:

$$\Gamma = \int_{H<U} dqdp, \quad S = k_B \ln \Gamma,$$

$$\Omega = \int_{U<H<U+\Delta} dqdp, \quad S' = k_B \ln \Omega.$$

Again give results for F, S, and U. (c) Explain why the two definitions give different results? (d) Apparently, the ensemble equivalence is violated here. Are you able to resolve the contradictions in (a) and (b) where the different ensembles and different definitions of entropy give different results [Hint: consider a set of N independent oscillators].

Problem 2.5. *Consider two classical non-interacting Hamiltonian systems with a total Hamiltonian $H = H_1 + H_2$. We define the phase space volume less than a given energy of each system by*

$$\Omega_j(E_j) = \int_{H_j<E_j} dp_jdq_j, \quad j = 1, 2,$$

and dp_jdq_j represents all the momenta and coordinates of the system j. (a) Give an expression for the phase space volume Ω of the combined system associated with H less than a given energy E in terms of $\Omega_1(E_1)$ and $\Omega_2(E_2)$. (b) Work in the canonical ensemble with the partition functions Z_j, show that entropy is additive, i.e., $S = S_1 + S_2$. (c) Comment on a similar statement of part (b) in the micro-canonical ensemble starting from the result of part (a) (i.e., is entropy additive in the micro-canonical ensemble?).

Problem 2.6. *Consider molecules moving in one dimension. The molecules are modeled as rigid rods of length a, they are confined between walls separated by a space of length L (much larger than a). The potential energy is 0 if the molecules do not overlap, and infinite if they overlap. The order of the molecules is maintained, i.e., they cannot pass through each other. (a) Calculate the canonical configuration partition function Q if there is only one molecule in the system. (b) Repeat the calculation if there are two molecules within the length L. (c) Generalize the results to system with an arbitrary number of N molecules. (d) Calculate the force exerted by the molecules on one of the walls for the case of one, two, and arbitrary N molecules.*

Problem 2.7. *Show that*

$$\langle (H, B(t)) A \rangle = k_B T \langle (A, B(t)) \rangle$$

where $A = A(p, q)$ and similarly for B are arbitrary classical dynamic variables which are zero outside a bounded region in phase space (Yvon's theorem). (A, B) stands for the Poisson bracket, and the average $\langle \cdots \rangle$ is with respect to the canonical distribution, $\rho = e^{-\beta H}/Z$. The time dependence means that $B(t) = B(p, q, t) = e^{t(\cdot, H)} B(p, q)$.

Problem 2.8. *Consider the generalized virial theorem and equipartition theorem with a classical Hamiltonian of the form*

$$H = \sum_{j=1}^{N} \frac{p_j^2}{2m_j} + V(q_1, q_2, \ldots, q_N).$$

Show that,

$$k_B T = \langle u \cdot \nabla H \rangle,$$

where the average $\langle \cdots \rangle$ is over the microcanonical ensemble. u is a $2N$-dimensional phase space vector of arbitrary function of $(q_1, q_2, \ldots, q_N, p_1, p_2, \ldots, p_N)$, such that $\nabla \cdot u = 1$, $\nabla = (\partial/\partial q_1, \ldots, \partial/\partial p_N)$ is the gradient operator in phase space. With a proper choice of u, show the validity of two important special cases of the general result:

$$k_B T = \left\langle \frac{p_j^2}{m_j} \right\rangle, \qquad k_B T = \left\langle q_j \frac{\partial H}{\partial q_j} \right\rangle.$$

Hint: use the Gibbs volume entropy as a starting point.

Problem 2.9.   *Consider the TPN (isothermal-isobaric) ensemble, where the temperature T, pressure p, and number of particles N are fixed fundamental thermodynamic variables. We define the partition function in the TPN ensemble as $Z_G(T, p, N) = \beta p \int_0^\infty dV \int \frac{d\Gamma}{N! h^{3N}} e^{-\beta(H+pV)}$, here the volume V is an integration variable, $\beta = 1/(k_B T)$, $d\Gamma = dq_1 dq_2 \cdots dq_{3N} dp_1 \cdots dp_{3N}$, H is the Hamiltonian, and h is the Planck constant. (a) Show that the Gibbs free energy is given by $G = U - TS + pV = -\frac{1}{\beta} \ln Z_G$. (b) Considering the ideal gas with $H = \sum_{j=1}^{3N} \frac{p_j^2}{2m}$, compute Z_G. You may need the definition of the Gamma function $\Gamma(x+1) \equiv x! = \int_0^\infty t^x e^{-t} dt$. (c) Using the thermodynamic relation, determine the chemical potential μ; and show that the ideal gas law, $pV = N k_B T$, is valid. (d) Show that the entropy S obtained from G agrees with the usual Sackur-Tetrode formula.*

Problem 2.10.   *According to Boltzmann, the total number of configurations in a volume V such that the i-th single particle phase space cell specified near $(\mathbf{r}_i, \mathbf{p}_i)$ is occupied by n_i particles is $W = \prod_i \frac{g_i^{n_i}}{n_i!}$. Here g_i is the degeneracy of the cell measured by phase space volume in units of h^3, that is, $g_i = d\mathbf{r} d\mathbf{p}/h^3$. With the additional constraints that the total number of particles is $N = \sum_i n_i$, and energy is $U = \sum_i n_i \mathbf{p}_i^2/(2m)$, show that entropy $S = k_B \ln W$ is given by the Sackur-Tetrode formula. Here the equilibrium distribution n_i is obtained by maximizing $\ln W$ subject to the two extra constraints.*

Chapter 3

Quantum Statistical Mechanics

It can be said that the origin of understanding entropy from a statistical mechanical point view gave birth to quantum mechanics, since the counting of states requires a discrete description in phase space by a unit h introduced by Planck. In the final version of quantum mechanics developed by Schrödinger, Heisenberg, Born, von Neumann, and others, we don't have phase space. Instead, we work in the Hilbert space. The classical formulation of statistical mechanics given in the last chapter needs only minor revision. All we need to do is to change the phase space integration, $\int \cdots d\Gamma/(N!h^{3N})$ to a Hilbert (or Fock) space trace, $\mathrm{Tr}(\cdots)$. In this chapter, we quickly give the fundamental equation, the von Neumann equation for the density matrix, and then consider one or two simple applications, such as Bose-Einstein condensation of free Bose particles.

3.1 Density matrix and von Neumann equation

In quantum mechanics, we cannot use phase space Γ, as positions and momenta are noncommutative quantities. In replacing the role of phase space, we work in Hilbert space [von Neumann (1955)], which is a very large space of all wave functions, $\Psi(q_1, q_2, \ldots)$. The coordinates $q = (q_1, q_2, \ldots)$ form the configuration space. An important feature of the Hilbert space is that it is a vector space over complex numbers, with an inner product (or scalar product) defined, satisfying the usual properties. The vector space builds in the superposition principle — if Ψ and Φ are wave functions, so is $\alpha\Psi + \beta\Phi$ — fundamental to quantum mechanics. The inner product associates a pair of vectors with a complex number, $(\Psi, \Phi) = \langle\Psi|\Phi\rangle$. The former notation is preferred by mathematicians, but the Dirac bra-ket notation is more common in the physics literature. Here we assume it is linear with respect to the second argument. We also have $(\Psi, \Phi) = (\Phi, \Psi)^*$, and we have a positive definite norm, $\|\Psi\|^2 = (\Psi, \Psi) > 0$ for $\Psi \neq 0$. The evolution of the system is governed by the Schrödinger equation

$$i\hbar\frac{\partial\Psi}{\partial t} = \hat{H}\Psi. \tag{3.1}$$

Here the quantum Hamiltonian is obtained from the classical Hamiltonian by the replacement, $p_j \to (\hbar/i)\partial/\partial q_j$.

Instead of the wave function, an equivalent description is the density matrix, $|\Psi\rangle\langle\Psi|$. This is a hermitian operator in the Hilbert space that produces a new wave function proportional to $|\Psi\rangle$, and is thus called the projector to the state Ψ. Such a density matrix represents a pure state. In statistical mechanics, we use the more general density matrix of mixed states [Landau and Lifshitz (1965)],

$$\hat{\rho} = \sum_i w_i|\Psi_i\rangle\langle\Psi_i|. \tag{3.2}$$

The addition of the states here is very much in the classical physics probabilistic sense, differing greatly from the wave function superposition that gives rise to interference effects. Here we only assume each wave function is normalized, $\langle\Psi_i|\Psi_i\rangle = 1$, and w_is represent discrete probabilities, such that $w_i > 0$, and $\sum_i w_i = 1$. It is easy to verify $\text{Tr}(\hat{\rho}) = 1$. And the operator $\hat{\rho}$ is positive semi-definite in the sense $\langle\varphi|\hat{\rho}|\varphi\rangle \geq 0$ for any wave function $|\varphi\rangle$. With the help of the density matrix, the quantum-mechanical and statistical averages can be combined as one of taking trace:

$$\langle A \rangle = \text{Tr}(\hat{\rho}A) = \sum_i w_i\langle\Psi_i|A|\Psi_i\rangle. \tag{3.3}$$

Here A is some arbitrary operator, and we have used the cyclic permutation property of trace, $\text{Tr}(ABC) = \text{Tr}(BCA)$.

We can give an equation of motion to $\hat{\rho}$, known as the von Neumann equation, starting from the Schrödinger equation. The basic assumption is that w_i is related to the initial 'preparation' of the system and does not change over time. Thus the change is only in the wave functions. The ket is given directly by the Schrödinger equation, (3.1), the but the bra, the vector in the dual space of ket, comes from taking the hermitian conjugate of the Schrödinger equation, $-i\hbar\partial\langle\Psi|/\partial t = \langle\Psi|\hat{H}$. We have assumed that the Hamiltonian is hermitian. After the differentiation, applying the Leibniz rule for product, we get

$$\dot{\rho} = \sum_i w_i \left(\frac{\partial |\Psi_i\rangle}{\partial t} \langle \Psi_i| + |\Psi_i\rangle \frac{\partial \langle \Psi_i|}{\partial t} \right)$$

$$= \frac{1}{i\hbar} \left(H\rho - \rho H \right) \equiv \frac{1}{i\hbar} [H, \rho]. \tag{3.4}$$

For notational simplicity, from now on we drop the hat symbol on the quantum-mechanical operators. In going from the first line to the second line, we note that H does not have an i index, and can thus be factored out. We can rewrite the summation back to ρ. This equation is structurally identical to the classical Liouville equation if we identify the commutator divided by $i\hbar$ with the Poisson bracket. We can solve the von Neumann equation in two ways, one sided, or two sided evolution:

$$\rho(t) = e^{[H, \cdot]t/(i\hbar)} \rho(0) = e^{-iHt/\hbar} \rho(0) e^{iHt/\hbar}. \tag{3.5}$$

Here the commutator with an empty slot is called a super-operator, since it operates on another operator, and is interpreted as $[H, \cdot]A = [H, A]$. Its effect is realized when the exponential function is Taylor expanded, $e^x = 1 + x + x^2/2! + \cdots$.

Since the von Neumann equation is formally the same as the Liouville equation, the argument leading to the micro-canonical, canonical and grand-canonical ensembles works exactly the same as before. The only change needed is to replace the phase space distribution function ρ by its quantum-mechanical counterpart, the density operator $\hat{\rho}$, and to replace the Hamiltonian function in phase space by the Hamiltonian operator. And finally, integration in phase space is to be replaced by quantum-mechanical trace. In particular, the classical partition function in the canonical ensemble is replaced by the trace of the exponential of the quantum Hamiltonian multiplied by $-\beta$,

$$Z = \int \frac{d\Gamma}{N! h^{3N}} e^{-\beta H} \quad \rightarrow \quad \text{Tr} \left(e^{-\beta \hat{H}} \right). \tag{3.6}$$

3.1.1 *Wigner transform*

It is possible to make a formal connection between the classical description based on the phase space distribution function ρ and the quantum density matrix $\hat{\rho}$. This connection is sort of fulfilled through the Wigner transform [Toda *et al.* (1992)], Chap. 1.3. For simplicity we consider a system of one degree of freedom. Classically, we describe the system with $\rho(q, p)$, and quantum-mechanically, with the density operator $\hat{\rho}$. Averages in the quantum case is computed by taking the trace. Trace is invariant with respect to the representation of the operator. If we represent $\hat{\rho}$ in the coordinate variables, the density matrix is $\langle x|\hat{\rho}|x'\rangle$. This is a function of two variables x and x', while classically it is in coordinate q and momentum p. Let us introduce a new pair of variables, known as Wigner transform,

$$q = \frac{x + x'}{2}, \quad r = x - x'. \tag{3.7}$$

We will identify the 'slow' variable q with the classical coordinate, and Fourier transform the 'fast' variable r into momentum space,

$$\rho_w(q,p) = \int_{-\infty}^{+\infty} dr\, e^{-ipr/\hbar} \left\langle q + \frac{r}{2} |\hat{\rho}| q - \frac{r}{2} \right\rangle. \tag{3.8}$$

By representing a physical observable $\hat{A}$ in the same way, we can demonstrate that

$$\begin{aligned}
\langle A \rangle &= \mathrm{Tr}\!\left(\hat{\rho}\hat{A}\right) \\
&= \int dx \int dx'\, \langle x|\hat{\rho}|x'\rangle \langle x'|\hat{A}|x\rangle \\
&= \int \frac{dq\,dp}{h} A_w(q,p)\rho_w(q,p).
\end{aligned} \tag{3.9}$$

Although this looks like a classical ensemble average with phase space counted in units of the Planck constant h, ρ_w is not guaranteed positive definite. A classical distribution emerges only after a suitable average in phase space and when $h \to 0$.

3.2   Examples of quantum systems

3.2.1   *Harmonic oscillators*

A system of coupled harmonic oscillators appears in many parts of physics — the vibration of solids, electromagnetic waves in a cavity, etc. Consider the Hamiltonian

$$H = \sum_{j=1}^{N} \frac{p_j^2}{2m_j} + \frac{1}{2} \sum_{i,j=1}^{N} K_{ij} q_i q_j. \tag{3.10}$$

Here m_j is the mass associated with degree j, and K the force constant matrix. We can eliminate the mass in the formula if we define new variables

$u_j = \sqrt{m_j}q_j$. Then the Hamiltonian can be written compactly in matrix notation,

$$H = \frac{1}{2}\dot{u}^T \dot{u} + \frac{1}{2}u^T \tilde{K}u, \qquad (3.11)$$

here u is a column vector, T denotes a matrix transpose, and the new force constant is normalized by mass, $\tilde{K}_{ij} = K_{ij}/\sqrt{m_i m_j}$. $\dot{u}$ is conjugate to u. The coupled oscillators can be diagonalized as a set of independent oscillators. Each of the new oscillators with a fixed oscillation frequency is known as a normal mode. Let a column vector Q with component Q_j be the so-called normal mode coordinates, we seek a linear relation $AQ = u$ such that $\tilde{K}$ is diagonalized, i.e.,

$$A^T \tilde{K} A = \{\omega_j^2\}. \qquad (3.12)$$

Here the right-hand side denotes a diagonal matrix with elements ω_j^2, which are the squares of the eigenfrequencies of the oscillations. For a real symmetric force constant matrix, $\tilde{K}$, it is possible to find a real and orthogonal matrix, $A^T A = I$, that diagonalizes $\tilde{K}$, where I is the identity matrix of dimension N. This choice makes the kinetic energy term simultaneously diagonal, thus we have,

$$H = \frac{1}{2}\dot{Q}^T \dot{Q} + \frac{1}{2}Q^T \{\omega_j^2\}Q = \sum_j \frac{1}{2}\left(\dot{Q}_j^2 + \omega_j^2 Q_j^2\right). \qquad (3.13)$$

The procedure of quantization is to change $P_j = \dot{Q}_j \rightarrow \frac{\hbar}{i}\partial/\partial Q_j$ as a differential operator, with the canonical commutation relation, $[Q_j, \dot{Q}_k] \approx i\hbar\delta_{jk}$. Introducing the ladder operators, $a_j = \sqrt{\omega_j/(2\hbar)}\,(Q_j + iP_j/\omega_j)$, the Hamiltonian can be written as $H = \sum_j \hbar\omega_j \left(a_j^\dagger a_j + \frac{1}{2}\right)$. The eigenvalues are well-known $\sum_j \hbar\omega_j \left(n_j + \frac{1}{2}\right)$, with $n_j = 0, 1, 2, \ldots$.

We now proceed to do statistical-mechanic calculations. Although it is possible to work in a microcanonical ensemble when all the oscillators have the same frequency, it is a difficult task to find the exact number of states with energy U in a given window. Thus, for the canonical ensemble, the problem is much easier. For independent oscillators, we simply have $Z = \prod_j z_j$, here

$$z_j = \sum_{n=0}^{\infty} e^{-\beta\hbar\omega_j(n+\frac{1}{2})} = \frac{1}{2\sinh(\beta\hbar\omega_j/2)}, \qquad (3.14)$$

is the partition function of a single oscillator with frequency ω_j. We have used a very useful summation formula, $1 + x + x^2 + x^3 + \cdots = 1/(1 - x)$, with $x = e^{-\beta\hbar\omega_j}$. The average total energy or the internal energy can be obtained by

$$U = -\frac{\partial \ln Z}{\partial \beta} = \sum_j \hbar\omega_j \left(\frac{1}{e^{\beta\hbar\omega_j} - 1} + \frac{1}{2} \right). \tag{3.15}$$

Here in the brackets the first term is the Bose function and the second term $1/2$ is called the zero-point motion contribution. The heat capacity of the oscillators is

$$C = \frac{dU}{dT} = k_B \sum_j \frac{y_j^2 e^{y_j}}{\left(e^{y_j} - 1 \right)^2}, \quad y_j = \beta\hbar\omega_j. \tag{3.16}$$

Einstein in 1906 proposed that solids are collections of quantum oscillators. For simplicity, he assumed that all the oscillators have the same frequency (which is approximately true only for the optical modes). Debye corrected this assumption, by considering the acoustic modes which have a continuous distribution of the frequencies down to 0. According to the Debye model, the heat capacity of a solid goes to 0 as T^3 which agrees better with experiments.

3.2.2 Quantum gases

For the oscillators example, the question of identical particles does not appear as each oscillator is distinct by its frequency. However, a collection of electrons, for example, has a new feature. A many-particle system of N identical particles must have a wave function that is either symmetric or anti-symmetric with respect to a permutation of the particles. This is because a swap of two particles does not lead to a new quantum state. But the wave function is defined only up to an arbitrary phase factor, i.e., Ψ and $\alpha\Psi$ with $|\alpha| = 1$ represent the same physical reality. In general, we say that quantum systems are described by rays in Hilbert space. Let us use an integer i to denote the i-th particle configuration $(\mathbf{r}_i, \sigma_i)$ for the position and z-component of spin. Then

$$\Psi(1, \ldots, i, \ldots, j, \ldots, N) = \alpha\Psi(1, \ldots, j, \ldots, i, \ldots, N). \tag{3.17}$$

Here we assume that this property of the wavefunction is universal and the phase factor $\alpha = e^{i\phi}$ is a constant independent of i and j. If we swap j and i on the right-hand side of the equation again, we get another α factor, but the wave function becomes the same as the original. Thus, $\alpha^2 = 1$. Two choices are possible, $\alpha = +1$ or -1. Particles that have symmetric wave functions with $\alpha = +1$ will be called bosons, and that have $\alpha = -1$ fermions. It can be shown from quantum field theory that fermions must have half integer spins, and bosons must have 0 or integer spins.

To enforce the symmetry or anti-symmetry using the wave functions is clumsy. We can build the basis of a many-particle quantum system with noninteracting single particle wave functions, say,

$$H^{(j)}\phi_r(j) = \epsilon_r \phi_r(j), \quad r = 0, 1, 2, \ldots. \tag{3.18}$$

Here integer j labels the particle number, and r labels the quantum state with ϵ_r as the eigenvalue of the single particle Hamiltonian. Taking the electrons of fermionic particles as an example, the wavefunctions must be antisymmetric. This can be achieved with the Slater determinant:

$$\Psi(1, 2, \ldots, N) = \frac{1}{\sqrt{N!}} \begin{vmatrix} \phi_{r_1}(1) & \phi_{r_2}(1) & \cdots & \phi_{r_N}(1) \\ \phi_{r_1}(2) & \phi_{r_2}(2) & \cdots & \phi_{r_N}(2) \\ \vdots & \vdots & \ddots & \vdots \\ \phi_{r_1}(N) & \phi_{r_2}(N) & \cdots & \phi_{r_N}(N) \end{vmatrix}. \tag{3.19}$$

Since a determinant is anti-symmetric with respect to a permutation of two rows, we have the required property, e.g., $\Psi(1, 2, \ldots, N) = -\Psi(2, 1, \ldots, N)$. Such many-body quantum state can be denoted abstractly as $|r_1 r_2 \cdots r_N\rangle$. Note that the order matters. In this representation, we list what the single particle states are, which are occupied by electrons. Alternatively, we can specify if a state is occupied or not by indicating its occupation number, n_r, as

$$|n_0, n_1 \cdots n_r \cdots\rangle. \tag{3.20}$$

Here the list is unending if the label r runs to infinity. For electrons, $n_r = 0, 1$ only, as a determinant with two identical columns produces a $\Psi = 0$ wave function which is a statement of non-existence in quantum mechanics (since the wave function should be normalizable to 1 to represent physical reality). This is the Pauli exclusion principle – you cannot put two electrons in the same state.

The many-particle states of all such forms of Eq. (3.20) are vectors in Fock space, which differs from usual elementary quantum mechanics in that the number of particles is variable. We can think that such states are created from the vacuum as [Fetter and Walecka (1971)]

$$|n_0, n_1 \cdots n_r \cdots\rangle = (c_0^\dagger)^{n_0} (c_1^\dagger)^{n_1} \cdots |0\rangle. \tag{3.21}$$

For electrons, the creation/annihilation operators satisfy

$$c_r c_s^\dagger + c_s^\dagger c_r = \delta_{rs}, \quad c_r c_s + c_s c_r = 0, \quad c_r^\dagger c_s^\dagger + c_s^\dagger c_r^\dagger = 0. \qquad (3.22)$$

For bosons, the creation and annihilation operators satisfy the algebra identical to the ladder operators for a set of harmonic oscillators. An important assertion of the creation/annihilation operator formulation is that Hamiltonians can be expressed solely in terms of these operators. For example, a system of electrons with Coulomb interactions hopping over a set of sites can be written as

$$\hat{H} = \sum_{ij} H_{ij} c_i^\dagger c_j + \frac{e^2}{2} \sum_{ij} c_i^\dagger c_j^\dagger v_{ij} c_j c_i, \qquad (3.23)$$

where the states are labelled by their site location i or j, and H_{ij} is the matrix element of a Hermitian matrix, $H = H^\dagger$, and $v_{ij} = 1/(4\pi\epsilon_0 r_{ij})$ is the Coulomb potential. We can diagonalize the first term, with eigenvalues ϵ_r, and the first term can be written as

$$\sum_r \epsilon_r c_r^\dagger c_r, \qquad (3.24)$$

but that will make the second Coulomb term more complicated.

The diagonal form is the same for fermonic or bosonic particles. The total number operator is $\hat{N} = \sum_r c_r^\dagger c_r$. The eigenvalues of the number operator are simply the sum, $\sum_r n_r$.

With this brief summary of the second quantization notation, we can calculate the grand-potential function for a noninteracting boson/fermion system of identical particles. The grand-partition function is

$$\Xi = \mathrm{Tr}\, e^{-\beta(\hat{H}-\mu\hat{N})} = \sum_\varphi \langle \varphi | e^{-\beta(\hat{H}-\mu\hat{N})} | \varphi \rangle$$

$$= \sum_{n_0, n_1, n_2, \ldots, n_r, \ldots} e^{-\beta(\sum_r \epsilon_r n_r - \mu \sum_r n_r)} = \prod_r \sum_{n_r} e^{-\beta(\epsilon_r n_r - \mu n_r)}$$

$$= \prod_r \left[1 + \eta e^{-\beta(\epsilon_r - \mu)} \right]^\eta. \qquad (3.25)$$

Here the trace is over any set of orthonormal many-particle states, which we have chosen to be these of the form of Eq. (3.21). Thus, they are the eigenstates of the Hamiltonian and total number operators. In the last step, we introduce $\eta = +1$ for fermions and -1 for bosons to combine the two results with a unified notation. In calculating the summation over n_r, we note for fermions of $\eta = +1$, we have only two terms, 1 and $e^{-\beta(\epsilon_r - \mu)}$, while

for bosons of $\eta = -1$, it is a geometric series 1, $e^{-\beta(\epsilon_r - \mu)}$, $\left(e^{-\beta(\epsilon_r - \mu)}\right)^2$, ...
The grand potential is

$$\Psi = U - TS - \mu N = -pV = -\frac{1}{\beta} \ln \Xi$$

$$= -\frac{1}{\beta} \sum_r \eta \ln\left(1 + \eta e^{-\beta(\epsilon_r - \mu)}\right). \tag{3.26}$$

Since $d\Psi = -SdT - pdV - \langle N \rangle d\mu$, we obtain the entropy as

$$S = -\frac{\partial \Psi}{\partial T} = \eta k_B \sum_r \ln(1 + \eta e^{-x_r}) + k_B \sum_r \frac{x_r e^{-x_r}}{1 + \eta e^{-x_r}}. \tag{3.27}$$

Here we have introduced a short-hand abbreviation, $x_r = \beta(\epsilon_r - \mu)$. This formula is not very illuminating. We will use the next formula,

$$\langle N \rangle = -\frac{\partial \Psi}{\partial \mu} = \sum_r \frac{1}{e^{x_r} + \eta}. \tag{3.28}$$

The physical meaning of this equation is fairly clear. Each term in the summation is just the average number of occupation, $\langle n_r \rangle = 1/(e^{\beta(\epsilon_r - \mu)} \pm 1)$, here plus sign is for fermions and minus sign for bosons. These are the Fermi-Dirac distribution and Bose-Einstein distribution, respectively. Finally, the equation of state is

$$pV = \eta k_B T \sum_r \ln(1 + \eta e^{-x_r}). \tag{3.29}$$

To get a more explicit equation of state, one has to eliminate the chemical potential μ in favor of the total average particle number $\langle N \rangle$ which cannot be done conveniently with simple elementary functions. Thus, we will just stop here.

Back to the entropy formula, we express x_r in terms of $\langle n_r \rangle$, as $x_r = \ln\left(\langle n_r \rangle^{-1} - \eta\right)$. The formula can be simplified further, after some algebra, given, for electrons or generally fermionic particles,

$$S(\eta = +1) = -k_B \sum_r \left[f_r \ln f_r + (1 - f_r) \ln(1 - f_r) \right], \tag{3.30}$$

here we have used $f_r = 1/(e^{\beta(\epsilon_r - \mu)} + 1)$ for the Fermi function. The above formula can be interpreted as the entropy mixing formula for the electrons with holes. For photon or phonons and bosonic particles,

$$S(\eta = -1) = k_B \sum_r \left[-N_r \ln N_r + (1 + N_r) \ln(1 + N_r) \right], \tag{3.31}$$

where $N_r = 1/(e^{\beta(\epsilon_r - \mu)} - 1)$ is the Bose-Einstein function. For both of these entropy formulas, we obtain Boltzmann's result if we omit the second term $(1 \pm \cdots)$. We can interpret the second term as a quantum correction. This also tells us that classical statistics works if the average occupation numbers are sufficiently small for all states r. In the supplementary reading materials, we will show how these same formulas can be obtained in a generalization along the lines of counting of Boltzmann to obtain entropy.

3.2.3 *Density of states*

The results obtained in the previous subsection are formal since we have not specified what their states r are and what the associated single particle energy ϵ_r is. In this subsection we assume that our particles are free particles each with Hamiltonian $\hat{\mathbf{p}}^2/(2m)$, confined in a box of side L and volume $V = L^3$. The confinement causes the wave vector or equivalently energy to be discretized. We apply the periodic boundary condition (we could also use fixed boundary conditions such that the wavefunction is 0 outside the box, but the periodic boundary condition is more convenient). The eigenstates of the Hamiltonian are then the plane waves,

$$\phi_r(\mathbf{r}) = \frac{1}{\sqrt{V}} e^{i\mathbf{p}_r \cdot \mathbf{r}/\hbar}, \tag{3.32}$$

with eigenvalues, $\epsilon_r = \mathbf{p}_r^2/(2m)$. Focusing on the x direction, the periodic boundary condition implies, $\phi(x, y, z) = \phi(x + L, y, z)$, thus we have the condition, $e^{ip_x L/\hbar} = 1$, or $p_x L/\hbar = 2\pi m_x$ with $m_x = \ldots, -1, 0, +1, \ldots$. The set of all possible values of momentum forms cubic lattice grids with lattice constant h/L, or $\mathbf{p}_r = (m_x, m_y, m_z)2\pi\hbar/L$.

In the thermodynamic limit, L will be macroscopically large. As a result, the lattice points become very dense. We can convert the discrete summation by integration with the following rule:

$$\sum_r I_r = \sum_{\mathbf{p}} I_{\mathbf{p}} (\Delta p)^3 \left(\frac{L}{h}\right)^3 \quad \rightarrow \quad \frac{V}{h^3} \int I_{\mathbf{p}} dp_x dp_y dp_z. \tag{3.33}$$

Here the integration is over the whole space of momentum. Further simplification is possible since the dispersion relation, $\epsilon = p^2/(2m)$, is isotropic, depending only on the magnitude of $\mathbf{p}$. Thus, the angular integration can be done immediately, and given as 4π, i.e., $dp_x dp_y dp_z = 4\pi p^2 dp$. We then make a change of variable from p to ϵ with $p\,dp/m = d\epsilon$. The final expression is then

$$\sum_r I_r = \int_0^\infty D(\epsilon)I(\epsilon)d\epsilon, \quad D(\epsilon) = \frac{V}{h^3}4\pi\,m^{3/2}\sqrt{2\epsilon}. \tag{3.34}$$

The function $D(\epsilon)$ is called the density of states.

The number of states per energy interval can also be obtained in the following way. Since each state in momentum space occupies a volume $(h/L)^3$, the total number of states $N(\epsilon)$ with energy less than ϵ is the volume of the sphere of radius p divided by h^3/V, and $\epsilon = p^2/(2m)$, so $N(\epsilon) = \frac{4\pi}{3}p^3/(h^3/V) = \frac{V}{h^3}\frac{4\pi}{3}(2m\epsilon)^{3/2}$. The density of states is then obtained by $D(\epsilon) = dN(\epsilon)/d\epsilon$.

With the concept of density of states, we can rewrite the grand-potential, average particle number, and internal energy as,

$$\Psi(T,V,\mu) = -\frac{\eta}{\beta}\int_0^\infty D(\epsilon)\ln\big(1 + \eta e^{-\beta(\epsilon-\mu)}\big)d\epsilon = -pV, \tag{3.35}$$

$$\langle N \rangle = \int_0^\infty \frac{D(\epsilon)}{e^{\beta(\epsilon-\mu)} + \eta}d\epsilon, \tag{3.36}$$

$$U = \int_0^\infty \frac{\epsilon D(\epsilon)}{e^{\beta(\epsilon-\mu)} + \eta}d\epsilon. \tag{3.37}$$

Here again $\eta = +1$ for fermions and -1 for bosons. With integration by parts, from the first equation, we can demonstrate for massive particles or classical, or quantum fermion or boson particles, $pV = \frac{2}{3}U$. Let $D(\epsilon) = a\sqrt{\epsilon}$, $a = 2^{5/2}\pi m^{3/2}V/h^3$, then $D(\epsilon)d\epsilon = \frac{2}{3}d(a\epsilon^{3/2})$, using the integration by parts formula, we find

$$
\begin{aligned}
pV &= \frac{\eta}{\beta}\int_0^\infty \ln\big(1 + \eta e^{-\beta(\epsilon-\mu)}\big)\frac{2}{3}d\left(a\epsilon^{3/2}\right) \\
&= \frac{2\eta}{3\beta}\left[a\epsilon^{3/2}\ln\big(1 + \eta e^{-\beta(\epsilon-\mu)}\big)\Big|_0^\infty - \int_0^\infty \frac{a\epsilon^{3/2}\eta e^{-\beta(\epsilon-\mu)}(-\beta)}{1 + \eta e^{-\beta(\epsilon-\mu)}}d\epsilon\right] \\
&= \frac{2}{3}\int_0^\infty \frac{\epsilon D(\epsilon)}{e^{\beta(\epsilon-\mu)} + \eta}d\epsilon = \frac{2}{3}U.
\end{aligned}
\tag{3.38}
$$

Here the first term in the second line evaluates to 0 in both limits. This result differs from the photon gas, which is $pV = U/3$.

3.2.4   *Electrons at low temperatures*

We apply the formulas in the previous subsection to calculate the heat capacity of electrons in a metal. Simple metals like Al can be described very well as free electrons. Electron carries a spin $1/2$, thus each orbital state is two-fold degenerate. We thus replace the density of states $D(\epsilon)$ by $2D(\epsilon)$. The total number of electrons and internal energy are

$$N = 2 \int_0^\infty f(\epsilon) D(\epsilon) d\epsilon, \quad U = 2 \int_0^\infty \epsilon f(\epsilon) D(\epsilon) d\epsilon. \qquad (3.39)$$

Here, $f(\epsilon) = 1/\big(\exp(\beta(\epsilon - \mu)) + 1\big)$ is the Fermi function. The heat capacity is obtained by taking the derivative of energy with respect to temperature,

$$C = \frac{dU}{dT} = 2 \int_0^\infty \epsilon D(\epsilon) \frac{df(\epsilon)}{dT} d\epsilon. \qquad (3.40)$$

However, the integrals involving the Fermi function cannot be performed in an elementary way. Fortunately, we can use a low temperature approximation. A typical distance between electrons is on the order of the lattice constant $a = 1\,\text{Å}$, thus the energy is of the order $\hbar^2/(2\,ma^2)$ — think of a particle in a box. This is some fraction of a hartree ($27.2\,\text{eV}$), which is much higher than the room temperature scale of order $k_B T \approx 25\,\text{meV}$. Thus lower temperature approximations even at $1000\,\text{K}$ are a good approximation. A series in temperature T known as Sommerfeld expansion can be made, [Ashcroft and Mermin (1976)], page 760,

$$\int_{-\infty}^{+\infty} \frac{H(\epsilon)}{e^{\beta(\epsilon - \mu)} + 1} d\epsilon = \int_{-\infty}^{\mu} H(\epsilon) d\epsilon + \frac{\pi^2}{6} \frac{1}{\beta^2} H'(\mu) + O\left(\frac{1}{(\beta\mu)^4}\right). \qquad (3.41)$$

This formula tells us that we can approximate the Fermi function as a step function, $\theta(\mu - \epsilon)$, and the correction is only second order in T. Thus, we have

$$N = \frac{16\sqrt{2}\pi}{3h^3} (m\epsilon_F)^{3/2} V + O(T^2), \quad \mu = \epsilon_F - \frac{\pi^2}{12} \frac{(k_B T)^2}{\epsilon_F} + O(T^4). \qquad (3.42)$$

To obtain the second chemical potential expression as a function of temperature, we need to maintain the first N expression to second order in T and then invert the equation. The Fermi energy ϵ_F is defined as $\mu(T \to 0) = \frac{1}{m} \left(\frac{3h^3 N}{16\sqrt{2}\pi V}\right)^{2/3}$, and the heat capacity is

$$C = \frac{\pi^2}{2} N \frac{k_B^2 T}{\epsilon_F}. \qquad (3.43)$$

Unlike the phonons in solids which has a cubic temperature dependence, the heat capacity of electrons is linear in T, and thus electrons make a larger contribution to heat capacity compared to phonons at low temperatures.

3.2.5 *Bose-Einstein condensation*

We now focus on systems which are bosonic, i.e., $\eta = -1$. According to the formula for $\langle N \rangle$, we must having $\mu < 0$, for otherwise we could get a negative particle number which is meaningless. In particular, the occupation number at the ground state (with ground state energy $\epsilon_0 = 0$) is

$$\langle n_0 \rangle = \frac{1}{e^{-\beta\mu} - 1} = \frac{1}{1 - \beta\mu + \cdots - 1} \approx \frac{1}{\beta(-\mu)}. \tag{3.44}$$

If the ground state does have a large number of particles of order N, then μ could be rather small, of order $1/N$. This is the phenomenon of Bose-Einstein condensation. In the thermodynamic limit, when $N \to \infty$, we can set the chemical potential to 0. However, our earlier formula for $\langle N \rangle$ is wrong by not taking into account this discontinuous behavior in the occupation numbers n_r. To remedy this, we separate out the ground state and the rest of the states, then we should have

$$\langle N \rangle = \langle n_0 \rangle + \int_0^\infty \frac{a\epsilon^{1/2}}{e^{\beta(\epsilon-\mu)} - 1} d\epsilon = \langle n_0 \rangle + bT^{3/2}, \tag{3.45}$$

$$b = a\, k_B^{3/2} \int_0^\infty \frac{x^{1/2}}{e^x - 1} dx \approx 2.31516\, a\, k_B^{3/2}. \tag{3.46}$$

The integral involving a Bose function times a power of x can be expressed by the Gamma function and the Riemann zeta function, as $\int_0^\infty x^{s-1}/(e^x - 1)dx = \Gamma(s)\zeta(s)$. Here we have given the numerical value at $s = 3/2$. We can write the result in the form, $\langle n_0 \rangle / \langle N \rangle = 1 - (T/T_c)^{3/2}$, see also Fig. 3.1. Here T_c is the transition temperature below which we have a finite fraction of the particles in the ground state. Above T_c, n_0 is of order 1, and $\langle n_0 \rangle / \langle N \rangle = 0$ in the thermodynamic limit. The formula, Eq. (3.45), is valid only below T_c.

It was thought that He^4 is a realization of the Bose-Einstein condensation phenomenon in nature. Helium-4 becomes a superfluid below $2.17\,K$. However, He^4 in the condensed phase is a fluid, not an ideal Bose-Einstein condensed gas. In the past decade or so, pure Bose-Einstein condensates were created by cooling down gas atoms by laser cooling to micro-kelvin range, such as rubidium, sodium, or lithium of alkali metals, which are

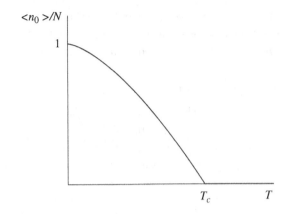

Fig. 3.1 The fraction of particles $\langle n_0 \rangle/N$ at the ground state vs. temperature T for ideal quantum Bose particles.

close to true ideal quantum Bose gas at low temperatures. We note that $T_c \propto (N/V)^{2/3}$. Thus, a gas phase (large volume) requires a low critical temperature.

3.3   Supplementary reading, Boltzmann counting method for quantum particles

For the purpose of counting states, we assume a single particle can have energy level ϵ_r with degeneracy g_r. We wish to put exactly n_r particles on the level r. For fermions, we can only place 0 or 1 particle in each state. Given g_r distinct states, we choose among them n_r being occupied and leave $g_r - n_r$ empty (we assume $g_r \gg n_r$). We can do this in $g_r!/\left[n_r!(g_r - n_r)!\right]$ ways for each level r. The total possible states given the set of choices of $\{n_r\}$ is [Pathria (1972)]

$$W_{\text{FD}} = \prod_r \frac{g_r!}{n_r!(g_r - n_r)!}. \tag{3.47}$$

Bose particles are somewhat harder to count. In this case, we are allowed to put any number of particles for each state, counting the degeneracies. We can think of each state as a box in which we put particles, which can be 0, 1, 2, ..., and there are g_r such boxes for each given r. The boxes are distinct, but the particles are not. We can align the particles and the walls between the boxes as objects and consider all possible permutations of them. This generates $(n_r + g_r - 1)!$ configurations. But permuting the particles or walls among themselves do not generate new states. Thus, we have

$$W_{\text{BE}} = \prod_r \frac{(n_r + g_r - 1)!}{n_r!(g_r - 1)!}. \tag{3.48}$$

The entropy expressions of the form in Eq. (3.30) and (3.31) are obtained, if we use the Boltzmann formula, $S = k_B \ln W$, and then the Stirling approximation. Since we have a g_r-fold degeneracy, we can identify with a non-degenerate notation r with $n_r = g_r f_r$ for fermions and $n_r = (g_r - 1)N_r$ for bosons, reproducing the earlier formula with an extra overall factor of g_r. For the classical Boltzmann case, the count W is given by Eq. (2.74). The equilibrium distribution for the occupation number n_r is obtained by maximizing the entropy subject to a fixed total energy and particle number, i.e., we seek n_i such that $\delta\left[\ln W - \beta(U - \mu N)\right] = 0$.

Problems

Problem 3.1. *Given the expression of a general density matrix as*

$$\rho = \sum_i w_i |\Psi_i\rangle\langle\Psi_i|, \quad w_i > 0, \quad \langle\Psi_i|\Psi_i\rangle = 1,$$

show that $\text{Tr}(\rho^2) \leq 1$. *And show that the equality holds only when the density matrix is a projector, i.e., $\rho^2 = \rho$.*

Problem 3.2. *The classical Poisson bracket is defined as*

$$(A, B) = \sum_j \left(\frac{\partial A}{\partial q_j} \frac{\partial B}{\partial p_j} - \frac{\partial B}{\partial q_j} \frac{\partial A}{\partial p_j} \right),$$

while the quantum mechanical version is defined by commutator

$$(\hat{A}, \hat{B}) = \frac{\hat{A}\hat{B} - \hat{B}\hat{A}}{i\hbar}.$$

Note that $(q_j, p_k) = \delta_{jk}$ for coordinates and conjugate momenta both in classical and quantum mechanics. Here δ_{jk} is the Kronecker delta which is 1 if $j = k$ and 0 otherwise. Show that the Poisson brackets have the following properties valid in both classical and quantum mechanics: (1) anti-symmetry $(A, B) = -(B, A)$; (2) linearity $(A + B, C) = (A, C) + (B, C)$, $(\lambda A, B) = \lambda(A, B)$, where λ is a constant; (3) Leibniz rule $(AB, C) = A(B, C) + (A, C)B$; and Jacob identity $((A, B), C) + ((B, C), A) + ((C, A), B) = 0$. This may require a lot of writing, be brief.

Problem 3.3. *Continue from the problem above, show that the canonical relation (Poisson bracket in classical mechanics and commutator in quantum mechanics), $(q_j, p_k) = \delta_{jk}$, is invariant with respect to the dynamics, i.e., it is the same whether p and q are at time $t = 0$ or $t > 0$. Show this explicitly. (a) for classical mechanics, let $A = Q_j(q, p, t)$, and $B = P_k(q, p, t)$, where Q_j and P_k are the solutions of the Hamiltonian equations of motion, with initial condition q and p at time $t = 0$, compute the Poisson bracket $(A, B)_{q,p}$, where the Poisson bracket is defined by the partial derivatives with respect to the initial positions and conjugate momenta. (b) For the quantum case, let $\hat{A}$ and $\hat{B}$ be the Heisenberg operators of positions and momenta, e.g., $\hat{A} = U(0, t)\hat{q}_j U(t, 0)$, where U is the unitary evolution operator. Compute the commutator $[\hat{A}, \hat{B}]$. (c) Argue or demonstrate explicitly that the conclusion is false if Q_j and P_k are at different times.*

Problem 3.4. *Consider N non-interacting atoms, each atom has three energy levels with energy $-\epsilon$, 0, and $+\epsilon$. (a) Determine the partition function Z of a canonical ensemble, and derive the associated entropy of the system S_1. (b) Now work in a micro-canonical ensemble. Work out a formula for the count $\Omega(E)$, which is the number of states with a total energy E of N atoms. Since energy is discrete in units of $\epsilon > 0$, the total energy E should be a multiple of ϵ. From the Boltzmann principle, determine the entropy S_2. (c) Show that the two entropies S_1 and S_2 calculated in the canonical and micro-canonical ensembles are equal in the thermodynamic limit.*

Problem 3.5. *Consider a general quantum system of identical particles with the Hamiltonian of the form*

$$H = \sum_{i=1}^{N} \frac{\mathbf{p}_i^2}{2m} + V = K + V,$$

where m is the mass, $\mathbf{p}_i$ is the momentum, and V is the potential energy operator that only depends on coordinates. Let the grand partition function be $\Xi = \mathrm{Tr}\exp\left[-\beta(H - \mu\hat{N})\right]$, where μ is chemical potential and $\hat{N}$ the number operator. Show that the average total kinetic energy K of the system can be expressed as $\langle K \rangle = m k_B T \partial \ln \Xi / \partial m$.

Problem 3.6. *Consider photons as particles having energy $\hbar\omega$ and chemical potential μ so that n such photons have energy $n\hbar\omega$. (a) Using the* *grand-canonical ensemble, compute the grand partition function* Ξ, *grand potential* Ψ, *entropy S, and average number of photons* $\langle N \rangle$. *(b) In fact, the chemical potential* μ *of a photon is 0. Give an argument as to why this is so. (c) Consider a collection of different modes (frequencies) of photons* $\omega_k = c2\pi k/L$, *where* $k = 1, 2, 3, \ldots$, *takes all the natural numbers, and c is the speed of light, and L is the length of a one-dimensional line on which the photons are constrained. Determine the (total) energy per unit length, u, of the thermal photons or black-body radiation, as a function of temperature T. You may assume that L is sufficiently large such that the summation can be turned into an integral.*

Problem 3.7. *Consider small vibrations of a one-dimensional chain of N atoms with the following Hamiltonian:*

$$H = \sum_{j=1}^{N} \frac{p_j^2}{2m} + \sum_{j=0}^{N} \frac{k}{2}(x_j - x_{j+1})^2, \quad x_0 = x_{N+1} = 0,$$

where, x_j is the displacement away from equilibrium and p_j is the associated conjugate momentum for site j, m is mass, and k is the force constant. Note that site 0 and N + 1 are fixed, and the dynamic equations apply only for the sites from 1 to N. (a) Compute the heat capacity of the classical system based on the equipartition theorem (in a canonical ensemble). (b) Assuming a solution of the form, $x_j = A\sin(jq)\cos(\omega_q t)$, determine the possible vibrational angular frequencies ω_q and wave number q. (c) Repeat the heat capacity calculation but for the corresponding quantum system.

Problem 3.8. *The eigenenergies of a single quantum particle in a one-dimensional box is $E_n = \frac{\pi^2\hbar^2 n^2}{2mL^2}$, where $n = 1, 2, 3, \ldots$, and m is mass, L is the length of the interval where the particle is confined. Working in a canonical ensemble: (a) Determine the force that the particle exerts to the wall of the box, in the high-temperature limit, i.e., $\beta = 1/(k_B T)$ is small. (b) Determine the same force, but in the opposite limit, i.e., temperature T is small. The answer should be accurate to the first order of a suitable small low-temperature expansion parameter.*

Problem 3.9. *Consider a quantum ideal gas of fermionic particles of spin one-half (e.g., electrons) in one dimension. The particles are confined in space from 0 to L with a periodic boundary condition. The energy of a single particle is $\epsilon_k = \hbar^2 k^2/(2m)$, $k = 2\pi l/L$, $l = \ldots, -2, -1, 0, 1, 2, \ldots$. (a) Derive the formula for the density of states $D(\epsilon)$, counting the spin* *degeneracy.* (*b*) *Express the grand partition function* Ξ *in terms of an integral involving* $D(\epsilon)$ *and chemical potential* μ. (*c*) *Compute the heat capacity* C *of the system of* N *particles in the low-temperature limit.*

Problem 3.10. *In the study of the phonon Hall effect, one needs to consider a Hamiltonian of the form*

$$H = \frac{1}{2}\left(p - Au\right)^{T}\left(p - Au\right) + \frac{1}{2}u^{T}Ku,$$

where matrix K *is symmetric and positive definite,* $A^{T} = -A$ *is antisymmetric,* u *is the mass normalized displacement vector and* p *is its conjugate momentum. Discuss the equations of motion, eigenmodes, and make a proposal to diagonalize such a system.*

Chapter 4

Phase Transitions, van der Waals Equation

In this chapter, we discuss equilibrium statistical mechanical problems where they cannot be solved exactly in most cases due to strong interactions. This is in contrast to the ideal gas for which the partition function can be computed exactly. A number of techniques have been developed over the years, such as series expansions over some small parameters, mean-field approximations, renormalization group theory, and computer simulations. We will study some of them.

4.1 Phase diagrams

A remarkable feature emerges when the particles are strongly interacting, that is, the occurrence of phase transitions. Phase transitions are common phenomena, for instance, H_2O molecules can exist in three phases, liquid, vapor, and ice. In reality, water can exist in the solid state in very many phases, see the diagram in Fig. 4.1. Two-dimensional water film is also very complex. All of these features are a consequence of intricate interactions that occur among the water molecules.

Ignoring the complexity of the solid phases, there are two distinguished points in the so-called phase diagram on the p-T plane. The triple point occurs at $T = 273.16\,\mathrm{K}$ and $p = 611.6\,\mathrm{Pa}$ (pascal). This is a unique point where ice, water, and vapor coexist. As we follow the line separating the vapor with liquid water, it turns out, the line ends at $T_c = 647\,\mathrm{K}$ and $p_c = 22.1\,\mathrm{Mpa}$. Beyond this point, there is no distinction between liquid and vapor phase at all; we call it fluid. This latter point is called the critical point. Near the critical point, light is strongly scattered by the fluid and

57

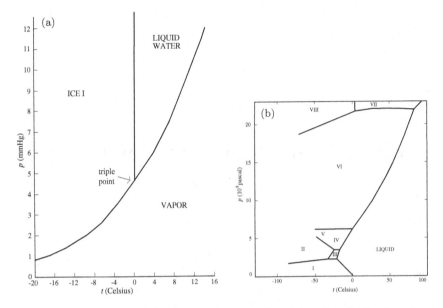

Fig. 4.1 (a) Water pressure vs. temperature near the triple point. (b) Ice in various phases at high pressure. Figures based on [Eisenberg and Kauzmann (1969)].

it is called critical opalescence. We will study the properties around this point in detail later.

When we change the system parameter such as temperature from low to high, passing through the liquid-vapor transition line, the system will absorb heat, $\Delta Q = T_0(S_g - S_l)$, where S_g is the entropy of the gas phase and S_l the liquid phase. This quantity is known as latent heat. Both the internal energy U and entropy S will have a discontinuous jump. Such a transition is classified as a first order phase transition. The slope of the transition line is given by the Clapeyron equation,

$$\frac{dp}{dT} = \frac{\Delta Q}{T\Delta V},$$ (4.1)

where ΔV is the volume change across the transition line. This equation is a simple consequence of the Gibbs-Duhem relation (see, [Callen (1985)], page 229–230).

Besides single component systems like water, other multiple component systems have more complicated phase diagrams. Here we show in Fig. 4.2 the mixture of water and methanol (a), and alloys of silver and copper (b). Liquid water can be mixed with liquid methanol at any mole fraction.

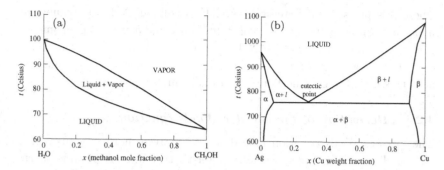

Fig. 4.2 (a) Water mixing with methanol at 1 atm. (b) Silver and copper alloy. Here α solid is silver rich and β solid is copper rich. The eutectic point has the lowest temperature of a pure liquid phase.

The same is true for the gas phase. However, there is a two-phase liquid and vapor coexistence region at which the mole fraction of liquid prefers more water, and vapor phase prefers more methanol. Various two-phase coexistence regions also exist in typical alloys. Gibbs gives a formula to determine the dimensionality of a phase, known as Gibbs phase rule,

$$f = r - M + 2, \tag{4.2}$$

where f gives the number of intensive variables that still can vary such that the system is still in that phase. r is the number of chemical components, M is the number of phases that are in coexistence. For example, at the triple point of water, $r = 1$, $M = 3$, so $f = 0$, nothing can vary. On the two-phase coexistence line of a simple fluid, $r = 1$, $M = 2$, then $f = 1$. The line is characterized by a one-dimensional intensive parameter (such as temperature T). For two component mixture such as water and methanol, $r = 2$.

4.2 The van der Waals equation

The critical point that terminates the distinction between liquid and gas was discovered by Thomas Andrews in the 1860s by experiments with carbon dioxide. Johannes D. van der Waals in his 1873 Ph.D. thesis proposed the first phenomenological theory to describe the behavior of such a non-ideal gas, now known as the van der Waals equation:

$$\left[p + a \left(\frac{N}{V} \right)^2 \right] (V - Nb) = N k_B T, \tag{4.3}$$

where p is pressure, T is temperature, V is volume, N is the number of molecules, k_B is the Boltzmann constant, and a and b are two phenomenological constants. He won the Nobel Prize in 1910 for this work. When $a = b = 0$, we recover the ideal gas law. The van der Waals equation describes both the gas and liquid states in a single equation.

4.2.1   Derivation of the van der Waals equation

The van der Waals equation indeed can be derived more rigorously in a suitable limit of very weak attractive inter-particle interactions and short-range repulsive hard particle exclusion [Lebowitz (1974)]. However, the math rigor required is beyond this course. Here we follow an intuitive but hand-waving argument to derive the Helmholtz free energy and study its consequences, [Domb (1996)], Chap. 2.

We work in the canonical ensemble. The partition function for N particles in a box of volume V is given by

$$Z = \frac{1}{N! h^{3N}} \int dq_1 dq_2 \cdots dq_{3N} \int dp_1 dp_2 \cdots dp_{3N} e^{-\beta H}. \qquad (4.4)$$

The Hamiltonian

$$H = \sum_{j=1}^{3N} \frac{p_j^2}{2m} + \sum_{i<j} v(r_{ij}), \qquad (4.5)$$

consists of the usual kinetic energy terms and interacting potential energy terms. The total potential energy is given by summing over pairs, once for each pair. The pair potential $v(r)$ only depends on the distance r between two particles, with a typical functional form, see Fig. 4.3, where it is repulsive at short distances and attractive at long distances. A well-known form is the Lennard-Jones potential, $v(r) = 4\epsilon \left(\left(\frac{\sigma}{r} \right)^{12} - \left(\frac{\sigma}{r} \right)^6 \right)$. This form of the inter-molecular interaction is an excellent model for noble gases, such as argon or krypton.

An exact evaluation of the partition function is not possible. We will make drastic approximations. First, when considering the q-integration, we focus on one atom, $\mathbf{r}$, assuming all the other atoms are fixed, ignoring the attractive potential, and assuming that molecules are hard spheres with some radius. That is, $v(r) = +\infty$ when $r < r_0$ and $v(r) = 0$ when $r > r_0$. Due to the volume exclusion, we get less volume available for the molecule,

$$\int d^3\mathbf{r}\, e^{-\beta \sum v(\cdot)} = V - bN. \qquad (4.6)$$

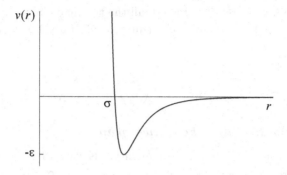

Fig. 4.3 Lennard-Jones pair interaction potential v vs. distance r.

Because of the reduced volume available to a molecule, the entropy is reduced. We use the same Sackur-Tetrode formula for the interacting molecular system with $V \to V - bN$:

$$S = Nk_B \ln(V - bN) + \frac{3}{2} Nk_B \ln T + \text{const.} \qquad (4.7)$$

We compute the Helmholtz free energy by the definition $F = U - TS$. For a mono-atomic ideal gas, the internal energy $U = \frac{3}{2} Nk_B T$. Due to the attractive interaction, the internal energy is also reduced. This is computed by

$$\left\langle \sum_{i<j} v(r_{ij}) \right\rangle = \frac{1}{2} N(N-1)\langle v \rangle \approx -a\left(\frac{N}{V}\right) N. \qquad (4.8)$$

In the above, we note that there are on the order of N^2 pairs. If we focus on one particular pair, say one at the origin, and the other at $\mathbf{r}$, we need to evaluate the average with respect to $\mathbf{r}$. The average is over the configurations weighted by the Boltzmann factor. We assume $\mathbf{r}$ is randomly distributed in the volume so the probability density is $1/V$. The average gets some nonzero value only when $\mathbf{r}$ is within some distance, so we can write $\langle v \rangle = \frac{\sigma^3}{V}\epsilon_0$. Lumping all the constants into a single a, we obtained the final form. Putting all the terms together, we have

$$F = -a\frac{N^2}{V} - Nk_B T \ln(V - bN) - C_V T \ln(T/e) + \text{const,} \qquad (4.9)$$

where $C_V = \frac{3}{2} Nk_B$ is the heat capacity at constant volume.

Once the free energy is obtained, all the thermodynamic properties are contained in it. For example, the equation of state, the van der Waals equation, is obtained by

$$p = -\frac{\partial F}{\partial V} = -a\left(\frac{N}{V}\right)^2 + \frac{Nk_BT}{V - Nb}. \tag{4.10}$$

4.2.2 Properties near the critical point

The general feature of the isothermal curves is shown in Fig. 4.4. At high temperatures, the system behaves as an ideal gas, $p \propto 1/V$. The pressure decreases with volume. As the temperature is lowered, a region of positive slope develops. This region signifies instability as pressure decreases as volume decreases instead of increases. The small volume and high pressure portion denotes liquid phase and large volume and lower pressure part denotes gas phase. The curve develops a minimum and a maximum at $\frac{\partial p}{\partial V} = 0$. There is a mid-point at which the second derivative changes sign. This point is called an inflection point mathematically. When the above mentioned three points meet at the same volume, critical point T_c is reached, above which the anomalous negative compressibility disappears, we are in a single fluid phase with no distinction between a liquid and gas.

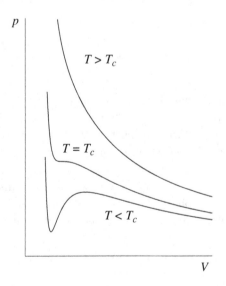

Fig. 4.4   The isotherms according to the van der Waals equation of state.

Thus, the condition for the critical point is

$$\frac{\partial p}{\partial V}\bigg|_T = 0, \qquad \frac{\partial^2 p}{\partial V^2}\bigg|_T = 0. \tag{4.11}$$

Using the van der Waals equation, we obtain two equations,

$$\frac{\partial p}{\partial V} = \frac{2aN^2}{V^3} - \frac{Nk_BT}{(V - Nb)^2} = 0, \tag{4.12}$$

$$\frac{\partial^2 p}{\partial V^2} = -\frac{6aN^2}{V^4} + \frac{2Nk_BT}{(V - Nb)^3} = 0. \tag{4.13}$$

Moving the second term to the right side and taking the ratio of the two equations, we can solve for the volume as $V = V_c = 3bN$. Putting this back to the first equation, we obtain $k_BT_c = \frac{8}{27}a/b$. And finally, using the equation of state, we find $p_c = a/(27b^2)$. These are the critical values determining the location of the critical point.

The isothermal compressibility is defined by

$$\kappa_T = -\frac{1}{V}\left(\frac{\partial V}{\partial p}\right)_T. \tag{4.14}$$

This quantity describes the percentage volume change when the pressure is increased. For a stable system, volume decreases when pressure increases, thus the minus sign. As we have noted, κ_T becomes negative below T_c in a certain volume range. Above T_c we can evaluate this quantity at $V = V_c$. we find

$$\kappa_T = \frac{4b}{3}\frac{1}{k_B(T - T_c)}. \tag{4.15}$$

As we can see, the isothermal compressibility diverges to infinity as the temperature approaches the critical value. In general, we define the critical exponent γ such that $\kappa \propto |T - T_c|^{-\gamma}$. For the van der Waals gas, $\gamma = 1$.

The critical isotherm is defined as the one passing through exactly the critical point. Thus, Taylor expanding the p-V curve around V_c at fixed T_c, we obtain

$$p(V, T_c) = p_c + \frac{\partial p}{\partial V}\bigg|_{V_c, T_c}(V - V_c) + \frac{1}{2}\frac{\partial^2 p}{\partial V^2}\bigg|_{V_c, T_c}(V - V_c)^2$$

$$+ \frac{1}{6}\frac{\partial^3 p}{\partial V^3}\bigg|_{V_c, T_c}(V - V_c)^3 + \cdots. \tag{4.16}$$

Since by definition at the critical point the first and second derivatives are 0, we have, at the critical isotherm $T = T_c$, $\Delta p = p - p_c \propto (V - V_c)^\delta$ with $\delta = 3$.

4.2.3 Maxwell's construction

Below the critical temperature T_c, we have the distinction of liquid phase and gas phase. In order to determine where the liquid phase ends and gas phase begins on the isotherm curve (as we increase the volume), Maxwell suggested a particular construction, which is to draw a horizontal line in such a way such that the areas enclosed by the horizontal line and the isotherm below and above the line are equal. This is based on the fact that there is one particular pressure (the horizontal line) for each given temperature $T < T_c$, such that there is a coexistence of liquid and gas phases. For $V < V_l$ we only see the liquid phase, and for $V > V_g$ we only see the gas phase. The coexistence is determined by the equality of chemical potentials, $\mu_l = \mu_g$.

The key assumption here is that the van der Waals equation is valid even in the unstable region and we use it to compute the chemical potential difference. Then the Gibbs-Duhem relation $SdT - Vdp + Nd\mu = 0$ implies, on the isotherm curve, $dT = 0$ (see Fig. 4.5),

$$N \int_1^3 d\mu = \int_1^3 Vdp = \int_1^A + \int_A^2 + \int_2^B + \int_B^3 = 0. \qquad (4.17)$$

Here we label the special points: 1 is the end point of liquid phase, 2 is in the unstable region, 3 is the beginning of gas phase, as we increase the

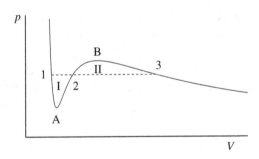

Fig. 4.5 The Maxwell construction of a van der Waals loop. The horizontal dotted line is drawn such that the areas of the regions I and II are the same.

volume. A and B are the minimum and maximum of the curve. The integration is interpreted as a line integral. The above condition is the same as $\mu_l = \mu(1) = \mu_g = \mu(3)$. Clearly, the four integrals can be regrouped and interpreted as the area bounded by the points 1-A-2-1 and 2-3-B-2 with algebraic sum 0. This is the Maxwell construction.

Based on the Maxwell construction, we can compute the volume change between the liquid and gas phase. Near the critical point, as shown by one of the problem set, we have

$$V_g - V_l \approx 4V_c\sqrt{\frac{T_c - T}{T_c}}, \qquad T < T_c. \tag{4.18}$$

Thus the distinction between liquid and gas phase diminishes in volume as $|T - T_c|^\beta$ with $\beta = 1/2$. This is another critical exponent, and we call it the order parameter critical exponent.

4.3 Series expansion to equation of state

Alternative to the approximate treatment in the van der Waals theory to the equation of state of a fluid is a series expansion, known as the virial expansion or cluster expansion [Reichl (1980); Friedman (1985); McQuarrie (2000); Hansen and I. R. McDonald (2006)]. The series expansion gives a rigorous description given the pair-interaction potential for the gas phase. However, a description of the liquid phase and the phase transition is not possible since we have only a finite number of terms in the series and the series stops to converge at high densities. Nevertheless, it is interesting to explore such methods since its technique has general validity for many other problems.

It is convenient to use the grand canonical ensemble to compute the first few terms of the grand partition function for small N, the number of particles, as

$$\Xi = \sum_{N=0}^{\infty} e^{\beta\mu N} Z_N, \quad Z_N = \frac{1}{N! h^{3N}} \int d\Gamma_N e^{-\beta H_N}, \tag{4.19}$$

here μ is chemical potential, Z_N is the partition function in the canonical ensemble of N particles. We define $Z_0 = 1$. $H_N = \sum_{i=1}^{N} \mathbf{p}_i^2/(2m) + v_N$ is the kinetic energy of N particles in three dimensions plus the interaction potential. The integration $d\Gamma_N$ is over all the components of momenta and all the coordinates. The momentum integrals are just Gaussian integrals, which can be performed analytically, given, for the grand partition function

$$\Xi = \sum_{N=0}^{\infty} \frac{x^N}{N!} Q_N, \quad x = \frac{e^{\beta\mu}}{\lambda^3}, \quad Q_N = \int d^3\mathbf{r}_1 d^3\mathbf{r}_2 \cdots d^3\mathbf{r}_N e^{-\beta v_N}. \quad (4.20)$$

The quantity $e^{\beta\mu}$ is known as fugacity, $\lambda = h/\sqrt{2\pi m k_B T}$ is the thermal wavelength, Q_N is the configurational partition function which only involves the coordinates of the atoms. We can easily obtain the first three coefficients of the expansion: $Q_0 \equiv 1$, $Q_1 = V$, here V is the volume occupied by the atoms. Since if we have only one atom, we do not have any "pair" interaction, $v_1 = 0$, the integration of $\mathbf{r}_1$ gives the total volume.

$$Q_2 = \int d^3\mathbf{r}_1 \int d^3\mathbf{r}_2 \, e^{-\beta v(|\mathbf{r}_1-\mathbf{r}_2|)}. \quad (4.21)$$

If we stop at Q_1 in the series expansion, we just recover the ideal gas law for the equation of state (show this). The Ursell and Mayer systematic expansion [Mayer and Mayer (1975)] introduces a small parameter f by

$$e^{-\beta v(r)} = 1 + f(r), \quad (4.22)$$

where $v(r)$ is the pair potential which depends only on the distance between two particles r. Then the general

$$Q_N = \int d^3\mathbf{r}_1 \cdots d^3\mathbf{r}_N \prod_{i<j}(1 + f_{ij}) \quad (4.23)$$

can be expressed in terms of diagrams. Each integration variable $\mathbf{r}_i$ is represented with a circle and a label, each $f_{ij} = f(|\mathbf{r}_i - \mathbf{r}_j|)$ factor is a line connecting the circles i with j. For example, for $N = 2$,

$$Q_2 = \int d^3\mathbf{r}_1 d^3\mathbf{r}_2 (1+f_{12}) = \circ \ \circ + \circ\!\!-\!\!\circ = V^2 + V \int dr 4\pi r^2 f(r). \quad (4.24)$$

The 3rd order term is, after expanding the $1 + f$ factors,

$$Q_3 = \int d^3\mathbf{r}_1 d^3\mathbf{r}_2 d^3\mathbf{r}_3 (1 + f_{12})(1 + f_{13})(1 + f_{23})$$

$$= \underset{\circ\ \ \circ}{\circ} + \underset{\circ\!-\!\circ}{\circ} + \cdots + \text{(diagrams)}$$

$$= V^3 + 3V^2 \int dr 4\pi r^2 f(r) + 3V \left[\int dr 4\pi r^2 f(r)\right]^2$$

$$+ \int d^3\mathbf{r}_1 d^3\mathbf{r}_2 d^3\mathbf{r}_3 f_{12} f_{23} f_{31}. \quad (4.25)$$

Putting the Q_N expressions in terms of the diagrams into Ξ, we obtain a Taylor expansion in variable x. Then the logarithm of Ξ is given by all diagrams that are connected, as

$$\ln \Xi = V \sum_{l=1}^{\infty} x^l b_l$$

$$= x \left[\circ \right] + \frac{1}{2!} x^2 \left[\circ\!-\!\circ \right] + \frac{1}{3!} x^3 \left[\diagup\!\!\diagdown + \bigwedge + \diagup\!\!\diagup + \triangle \right] + \cdots .$$

$$(4.26)$$

The pressure can be calculated with the general thermodynamic relation $\Psi = U - TS - \mu N = -pV = -k_B T \ln \Xi$. Or $p = \frac{k_B T}{V} \ln \Xi$. However, this is not the 'equation of state', as it is in variable x. For the equation of state we need to express p in terms of V, N, and T. To eliminate the chemical potential μ in favor of the particle number, or density, we use

$$\rho = \frac{N}{V} = \frac{1}{V} \frac{\partial \ln \Xi}{\partial (\beta \mu)} = \sum_{l=1}^{\infty} l x^l b_l. \qquad (4.27)$$

Note that $x = e^{\beta \mu}/\lambda^3$, thus $\partial x^l / \partial (\beta \mu) = l x^l$. We can eliminate x in favor of ρ. This is obtained by inverting the series for ρ, as $x = \rho - 2b_2 \rho^2 + O(\rho^3)$, and substituting x into p. We can express the result as

$$p = k_B T \sum_{l=1}^{\infty} a_l(T) \rho^l, \qquad (4.28)$$

where the virial coefficients a_l can be expressed by a class of more compact diagrams known as star graphs (or doubly connected graphs). The first few terms are

$$a_1 = 1, \qquad (4.29)$$

$$a_2 = -\frac{1}{2} \frac{1}{V} \left[\circ\!-\!\circ \right], \qquad (4.30)$$

$$a_3 = -\frac{1}{3} \frac{1}{V} \left[\triangle \right], \qquad (4.31)$$

$$a_4 = -\frac{1}{4 \cdot 2} \frac{1}{V} \left[3 \, \square + 6 \, \boxtimes + \boxtimes \right]. \qquad (4.32)$$

In general, a_l has an overall numerical factor of $-\frac{1}{l} \frac{1}{(l-2)!} \frac{1}{V}$, and each graph may have an extra integer factor related to the symmetry of the graph.

Problems

Problem 4.1. *The Gibbs phase rule determines the degrees of freedom of intensive parameters in a given phase or phase coexistence region, by* $f = r - M + 2$, *where r is the number of chemical components, M is the number of coexistence phases. Derive this result using the Gibbs-Duhem relations, one for each phases.*

Problem 4.2 (Lever rule). *In an experimental p-V isotherm curve there is a flat plateau for temperature below T_c. If the maximum volume of a pure liquid phase is V_l and the minimum volume of a pure gas is V_g, and a mixed two-phase coexistence has volume V, with a fixed total particle number N, show that the mole fractions of liquid and gas are,*

$$x_l = \frac{N_l}{N} = \frac{V_g - V}{V_g - V_l}, \quad x_g = \frac{N_g}{N} = \frac{V - V_l}{V_g - V_l},$$

respectively ($V_l < V < V_g$ are the left end, a point in-between, and right end of the flat portion of the p-V isotherm. See also [Callen (1985)], page 239).

Problem 4.3. *Given the van der Waals equation:*

$$p = -a\frac{N^2}{V^2} + \frac{Nk_BT}{V - Nb},$$

the Maxwell construction can also be expressed as

$$\int_{V_l}^{V_g} \left(p - \bar{p}\right)dV = f(V_g) - f(V_l) = 0,$$

where $\bar{p}$ is the value of the pressure corresponding to the cut in a Maxwell construction (the coexistence pressure in a two-phase region). Note that $\bar{p}$ also satisfies the van der Waals equation when $V = V_l$ or V_g. (a) Find the function $f(V)$ of volume V. (b) Using the result in (a), assuming that the liquid-gas coexistence curve is a symmetric function around the critical value V_c, $V_g = V_c + x$, $V_l = V_c - x$, and $V_g - V_l = 2x$, show that

$$x \approx 2V_c\sqrt{(T_c - T)/T_c}, \quad T < T_c,$$

in the asymptotic critical region when T is close to T_c (x is small).

Problem 4.4. (*a*) *Derive the Helmholtz free energy of the van der Waals theory for fluid (hand-waving arguments are acceptable):*

$$F = -a\frac{N^2}{V} - Nk_BT\ln(V - bN) - \frac{3}{2}Nk_BT\ln(T/c) + \text{const.}$$

*The constants a, b, c, const, are independent of both volume V and temperature T. (*b*) Calculate the heat capacity at constant volume, C_V. (*c*) Calculate the heat capacity at constant pressure, C_p. (*d*) Give the asymptotic expression as a function of temperature T for C_p near the critical point at a fixed critical pressure p_c.*

Problem 4.5. *Consider the equation of state of a classical monoatomic gas slightly modified from the ideal gas law. We assume a pair-wise interaction potential $v(r)$ between two particles with the distance $r = |\mathbf{r}|$ dependence as*

$$v(r) = \begin{cases} \infty, & r < a, \\ -\epsilon, & a \leq r < b, \\ 0, & r \geq b. \end{cases}$$

(*a*) *Write down the partition function Z_N of N particles in a canonical ensemble and express it as a product of two factors, a temperature dependent part and a configurational partition function, $Z_N = P_N Q_N$. Evaluate the temperature dependent part, P_N. (*b*) Give the expression for the grand partition function Ξ of the grand-canonical ensemble, using the result in (*a*). Evaluate the configurational partition functions in a box of volume V much larger than a^3 and b^3 for zero particle Q_0, one particle Q_1, and two particles Q_2. (*c*) Determine the equation of state (pressure p as a function of volume V and temperature T), using the results in (*b*) above, based on Q_0, Q_1, Q_2, ignoring the higher order terms Q_N for $N > 2$ for the grand partition function Ξ.*

Problem 4.6. *In the virial series expansion of the equation of state, it is found that the configuration partition function of N particles, Q_N, when expanded in the small parameter $f = e^{-\beta v} - 1$, contains both connected and disconnected graphs. Here disconnected graphs mean that the integral can be written as a product of two or more factors. (*a*) Show explicitly by Taylor expanding the logarithm, that $\ln \Xi = \ln(1 + xQ_1 + x^2Q_2/2 + x^3Q_3/6 + \cdots)$, to 3rd order in x, contains only the connected graphs. (*b*) Present an argument that this is true in general (hint: the grand potential $\Psi = -k_BT\ln\Xi$ must be an homogeneous function of degree 1 in the volume in the thermodynamic limit).*

Chapter 5

Ising Models and Mean-Field Theories

Magnetism is truly a quantum phenomenon as classical charged systems cannot support a permanent magnet in equilibrium due to the Bohr-van Leeuwen theorem. This theorem states that the equilibrium average of magnetization $\langle \mathbf{M} \rangle$ for a classical statistical-mechanical system is 0. Consider a general collection of particles each with charge q_i and mass m_i, the Hamiltonian is

$$H = \sum_{i=1}^{N} \left[\frac{1}{2m_i} \left(\mathbf{P}_i - q_i \mathbf{A}(\mathbf{r}_i) \right)^2 + q_i \phi(\mathbf{r}_i) \right] + U(\mathbf{r}_1, \ldots, \mathbf{r}_N). \tag{5.1}$$

Here $\mathbf{A}$ is the vector potential and ϕ the scalar potential. U is the interacting potential energy independent of the momenta. The magnetization is related to the (gauge invariant) angular momentum $\mathbf{l}_i = \mathbf{r}_i \times m_i \mathbf{v}_i$. Here the velocity is related to the conjugate momentum by $\mathbf{P}_i = m_i \mathbf{v}_i + q_i \mathbf{A}_i$. After a change of integration variables from $\mathbf{P}_i$ to $\mathbf{v}_i$, we can eliminate $\mathbf{A}$. The Hamiltonian is an even function of $\mathbf{v}_i$ and the angular momentum is odd in $\mathbf{v}_i$, the integration over $\mathbf{v}_i$ is zero by symmetry. Thus, the average in canonical ensemble is $\langle \mathbf{l}_i \rangle = 0$. Alternatively, one can compute the partition function. By a similar change of variables, one can see that the partition function is independent of the vector potential, and thus, the magnetic field. Since the derivative of the Helmholtz free energy with respect to the magnetic field is the magnetization, we obtain again 0 magnetization.

5.1   The Ising model

Classically the magnetic dipole moment is

$$\mathbf{M} = \frac{1}{2} \int \mathbf{r} \times \mathbf{j} \, dV = I \times (\text{area}), \qquad (5.2)$$

where $\mathbf{j}$ is the current density vector. If there is a current I running through a loop enclosing an area A, the magnitude of the magnetic moment is given by the product of the two with a direction normal to the area. The vector potential produced by such a loop current is

$$\mathbf{A} = \frac{\mu_0}{4\pi} \frac{\mathbf{M} \times \mathbf{r}}{r^3}, \qquad (5.3)$$

in SI units. The magnetic induction is given by $\mathbf{B} = \nabla \times \mathbf{A}$. The energy of the magnetic moment in a field $\mathbf{B}$ is $E = -\mathbf{M} \cdot \mathbf{B}$.

Quantum-mechanically, for an atom, unpaired electron has spin $\mathbf{s}$ and the electron orbital motion generates angular momentum $\mathbf{L} = \mathbf{r} \times \mathbf{p}$ with the total angular momentum $\mathbf{J} = \mathbf{L} + \mathbf{s}$. The magnetic moment of an atom is

$$\mathbf{M} = -g\mu_B \mathbf{J}/\hbar, \quad \mu_B = \frac{e\hbar}{2m_e}, \qquad (5.4)$$

where g is known as the Landé factor, μ_B is the Bohr magneton, m_e is the mass of an electron. $e > 0$ is the unit of charge, and the minus sign is because electrons carry negative charge. $\mathbf{J}$ is a vector quantum operator, satisfying the angular moment commutation relations, e.g., $[J_x, J_y] = i\hbar J_z$.

We consider a simple case where the total angular momentum is contributed only by electron spin of $\mathbf{s} = \frac{\hbar}{2}\vec{\sigma}$, here the components of the vector Pauli matrices are

$$\sigma^x = \begin{pmatrix} 0 & 1 \\ 1 & 0 \end{pmatrix}, \quad \sigma^y = \begin{pmatrix} 0 & -i \\ i & 0 \end{pmatrix}, \quad \sigma^z = \begin{pmatrix} 1 & 0 \\ 0 & -1 \end{pmatrix}. \qquad (5.5)$$

If we put a collection of (classical) magnetic moments together, the interaction between each pair has energy $U = \frac{\mu_0}{4\pi r^3}\left[\mathbf{M} \cdot \mathbf{M} - 3(\mathbf{M} \cdot \mathbf{r})(\mathbf{M} \cdot \mathbf{r})\right]$. If one plugs in typical values of atomic parameters, this give $U \approx 10^{-4}\,\text{eV}$, too small for ferromagnetism near room temperature. The origin of room temperature magnetization is not due to magnetic dipole interactions; rather, it is truly a quantum effect of exchange-correlation interaction of the electrons. Heisenberg proposed such a model based on many-body quantum mechanics, giving the Hamiltonian as

$$H = -J \sum_{\langle i,j \rangle} \vec{\sigma}_i \cdot \vec{\sigma}_j + g\mu_B B \sum_i \sigma_i^z. \tag{5.6}$$

Here we assume that the magnetic field is along the positive z-direction. The summation $\langle i, j \rangle$ is over the nearest neighbor pairs, only once. The first term here is rotationally symmetric. In a crystal it may happen that the three directions may not be equivalent, giving rise to the anisotropic Heisenberg model terms $-J_x \sigma_i^x \sigma_j^x - J_y \sigma_i^y \sigma_j^y - J_z \sigma_i^z \sigma_j^z$. If $J_z = 0$, it is called an XY model. The Ising model takes only the σ^z components, with $J_x = J_y = 0$. Since all the z-component Pauli spin matrices commute with each other, the Ising model Hamiltonian is not a quantum operator but just a c-number:

$$E = -J \sum_{\langle i,j \rangle} \sigma_i \sigma_j - h \sum_i \sigma_i, \quad \sigma_i = \pm 1. \tag{5.7}$$

Here we have defined $h = -g\mu_B B$ as our 'magnetic field' for the Ising model. The Ising model was proposed by Wilhelm Lenz around 1920s, to his student Ernst Ising. He was able to solve the problem in one dimension and found no phase transitions. A two-dimensional square lattice problem was solved exactly by Lars Onsager much later in 1944. Although we cannot solve it in three dimensions, the Ising model is simple enough for us to try various approximation methods, and is thus a very good model to discuss phase transitions and some of the conceptual issues for many-body interacting systems.

5.1.1   *Paramagnetism*

We consider a particularly simple case, the paramagnetic Ising model. We can say this is like the "ideal gas" of magnetism. The model is obtained when the spin-spin coupling J is set to 0. Thus,

$$H = -h \sum_i \sigma_i. \tag{5.8}$$

Since the energy is a sum of independent terms, the partition function can be computed by focusing only on one spin, so $Z = z^N$, with

$$z = \sum_{\sigma = \pm 1} e^{-\beta(-h\sigma)} = e^{\beta h} + e^{-\beta h}. \tag{5.9}$$

The free energy is given by

$$F = -\frac{1}{\beta} \ln Z = -k_B T N \ln \left(2 \cdot \cosh \frac{h}{k_B T} \right). \tag{5.10}$$

The above Helmholtz free energy completely characterizes the thermodynamic properties. All other thermodynamic quantities can be derived from it. Remembering that $dF = -SdT - \cdots$, the entropy is given by

$$S = -\frac{\partial F}{\partial T} = k_B N \ln \left(2 \cdot \cosh \frac{h}{k_B T} \right) - \frac{Nh}{T} \tanh \frac{h}{k_B T}. \tag{5.11}$$

This formula looks rather ugly. We can put it in a more appealing form (see the problem set for this chapter), as the Gibbs entropy of mixing,

$$S = k_B N \left(-p_+ \ln p_+ - p_- \ln p_- \right), \quad p_+ = \frac{1+m}{2}, \quad p_- = \frac{1-m}{2}, \tag{5.12}$$

here p_+ is the probability of the spin being 'up' ($\sigma = 1$) and p_- the probability of the spin being 'down' ($\sigma = -1$). The magnetization per spin, $m = M/N$ is obtained below:

$$M \equiv \left\langle \sum_i \sigma_i \right\rangle = -\frac{\partial F}{\partial h} = N \tanh(\beta h). \tag{5.13}$$

From the above equation, we can obtain the full differential of the free energy

$$dF = \frac{\partial F}{\partial T} dT + \frac{\partial F}{\partial h} dh = -SdT - Mdh. \tag{5.14}$$

This is the fundamental thermodynamic differential analogous to $dU = TdS - pdV + \mu dN$ for the gas system. We also see that F should be considered as a function of two variables T and h, both of them intensive variables. Thus, it is more like the Gibbs free energy G. But it is too late to change the name, and we will call it Helmholtz free energy because it is obtained by $F = -k_B T \ln Z$.

The internal energy U can be computed by $U = F + TS$. Alternatively, one can obtain the internal energy as an average of the total Hamiltonian, $U = \langle H \rangle = -\partial \ln Z / \partial \beta = -hN \tanh(\beta h) = -hM$. The magnetic susceptibility is defined by

$$\chi = \left. \frac{\partial M}{\partial h} \right|_{h \to 0} = \left. \frac{\beta N}{\cosh^2(\beta h)} \right|_{h=0} = \frac{N}{k_B T}. \tag{5.15}$$

The behavior that the susceptibility goes as the inverse of temperature is known as Curie's law.

Although the model we have here is very simple, the thermodynamic relations are of general validity, such as $dF = -SdT - Mdh$, or $U = -\partial \ln Z/\partial \beta$. In the next section, we discuss the mean-field theory where spin-spin interactions make the problem not solvable exactly. We have to resort to approximations.

5.2 Mean-field theory

Due to the couplings between the spins, the exact computation of the partition function becomes a rather nontrivial problem. The strategy of the mean-field approach is to reduce the many-body problem back to a 'single body' problem, by focusing on only one spin. We will take this spin as at the site 0 surrounded by its $q = 2d$ neighbors, in a d-dimensional hypercubic lattice. However, in order to make it a one-body problem, the spins of the neighbors will be replaced by their average values, $\sigma_j \to \langle \sigma_j \rangle$. They are no longer 'dynamic variables'. The Hamiltonian for the spin at site 0 is then given by

$$H_0 = -J \sum_j \sigma_0 \langle \sigma_j \rangle - h\sigma_0 = -(Jqm + h)\sigma_0. \tag{5.16}$$

Here we assume that each site is completely equivalent to other sites, thus $\langle \sigma_j \rangle = m$ is independent of the site index j. In the above equation, h is the explicit field, while Jqm is an additional field produced by the surrounding spins. It is called the 'mean field'. We can calculate the average magnetization of the spin at site 0. However, it should be the same as the average value of the other spins due to the complete translational symmetry of the lattice. Thus, we obtain the mean-field equation

$$m = \langle \sigma_0 \rangle = \tanh\Big(\beta(Jqm + h)\Big). \tag{5.17}$$

This result is obtained if we compare the present problem with the paramagnet problem of Eq. (5.13). As the magnetization appears on both the left and right side of the equal sign, this is indeed an equation for m. We can solve the equation formally by moving the magnetization m to one side, using the inverse tanh function, to obtain

$$h = -mqJ + \frac{1}{\beta}\text{arctanh}(m), \tag{5.18}$$

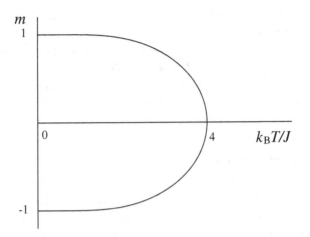

Fig. 5.1 The mean-field result of magnetization m vs. dimensionless temperature $k_B T/J$ in two dimensions ($q = 4$) at $h = 0$.

where $\text{arctanh}(m) = \tanh^{-1}(m) = \frac{1}{2}\ln\frac{1+m}{1-m}$, valid only when $|m| < 1$. This equation gives an h vs. m curve for fixed temperature $T = 1/(k_B\beta)$.

For the particular case of zero magnetic field, $h = 0$, we can solve β in terms of m, as

$$\beta = \frac{1}{k_B T} = \frac{1}{2qJm}\ln\frac{1+m}{1-m}, \quad |m| < 1. \tag{5.19}$$

This gives us a β vs. m plot, but we can plot m vs. T instead, see Fig. 5.1. A striking feature of this plot is the occurrence of a 'spontaneous magnetization', where we get a non-zero m when we do not have an applied external magnetic field, provided that the temperature is below a certain value T_c. This temperature is called the critical temperature, below which we have ferromagnetic phases and above which we are in a paramagnetic phase. There is a phase transition when one passes the critical point. This transition is termed a second order phase transition.

We now analyze the functional behavior of physical quantities near the critical point. First, we need to determine the critical temperature T_c. From Eq. (5.19), we see that the expression $\frac{1}{2m}\ln\frac{1+m}{1-m}$ has numerical value larger than 1, when $0 < m < 1$. When m is close to 0, we can make a Taylor expansion with m and find that the value approaches 1 as m approaches 0. When m is close to 1, the denominator $(1 - m)$ makes the value rather large. This means, in order to have a nonzero m, we have to have $\beta qJ > 1$.

Table 5.1 Comparison of mean-field and exact critical value $k_B T_c / J$ for hypercubic lattices in d dimensions.

d	1	2	3	4	5
MF	2	4	6	8	10
exact	0	2.27	4.51	6.68	8.78

The critical temperature is obtained by the limit $m \to 0$, or when

$$\beta_c J q = 1 \quad \text{or} \quad T_c = \frac{Jq}{k_B}. \tag{5.20}$$

How good is this mean-field prediction for a d-dimensional hypercubic lattice? In Table 5.1, we show the mean-field value of $k_B T_c / J$ to that of exact (or numerical) results. As we can see, the mean-field prediction is completely wrong in one dimension, since the mean-field result predicted a finite critical temperature but the actual 1D model has $T_c = 0$. The two-dimensional square lattice has an exact value of $2.269\ldots$, the mean-field result is 4, off by about a factor of 2. However, as the dimensionality goes higher, the mean-field approximation becomes better. This is a very general feature of the mean-field theory.

Knowing the value of T_c we can study how the magnetization approaches 0 as temperature approaches T_c. Setting $h = 0$ in Eq. (5.17), and then expanding the hyperbolic tangent function according to $\tanh(x) = x - \frac{1}{3}x^3 + \cdots$, we find, for small m,

$$m = \beta q J m - \frac{1}{3}(\beta q J m)^3 + \cdots \quad \to \quad m^2 \approx \frac{3}{(\beta q J)^3}(\beta q J - 1). \tag{5.21}$$

Clearly, $m = 0$ is also a possible solution. This corresponds to the magnetization at $T > T_c$. If $m \neq 0$, we can cancel m, and arrive at the equation after the arrow. Noting that $\beta_c q J = 1$, we can replace 1 by $\beta_c q J$ and find $m^2 \propto \beta - \beta_c$, or

$$m \propto (T_c - T)^{1/2}, \quad T \to T_c. \tag{5.22}$$

This gives the mean-field theory prediction that order parameter m approaches 0 as $1/2$ power of the temperature difference. Notice that this is the same value as that of the van der Waals gas for the density (or volume) difference.

The magnetic susceptibility at zero field, defined as the rate of change of magnetization with respect to the magnetic field evaluated at $h = 0$, measures the fluctuations of the magnetization. By taking the derivative with respect to h on both sides of the mean-field equation, Eq. (5.17), we obtain

$$\chi = \frac{\partial m}{\partial h} = \frac{1}{\cosh^2\left(\beta(qJm + h)\right)} \left(\beta qJ\frac{\partial m}{\partial h} + \beta\right). \qquad (5.23)$$

This is still an equation for χ. The solution will be different depending on if T is below or above T_c. Consider the simpler case of $T > T_c$, then $m = 0$ when $h \to 0$, we find, at zero field, $\chi = \beta qJ\chi + \beta$, or

$$\chi = \frac{1}{k_B} \frac{1}{(T - T_c)}, \quad T > T_c. \qquad (5.24)$$

We have used the fact that $\beta_c qJ = qJ/(k_BT_c) = 1$. We see that the susceptibility diverges as $|T - T_c|^{-\gamma}$ with $\gamma = 1$. Such a divergent behavior is known as the Curie-Weiss law.

Figure 5.2 shows the isotherms, m vs. h for $T > T_c$, $T = T_c$, and $T < T_c$. The asymptotic critical isotherm (m small) can be obtained by Taylor expanding the right-hand side of the mean-field equation, setting $\beta = \beta_c$, we find $m = \tanh(m + \beta_c h) \approx m + \beta_c h - \frac{1}{3}(m + \beta_c h)^3 + \cdots$. After canceling m, we find $\beta_c h \approx \frac{1}{3}m^3$, or $h \sim m^\delta$ with $\delta = 3$.

For $T < T_c$ we see the typical "van der Waals" loop if we flip the axes as h vs. m, so that h is analogous to pressure and m to the volume difference. The same Maxwell construction is needed in order to represent equilibrium thermodynamic behavior, but in this case it is very simple, one jumps from the positive magnetization $+m_0$ to $-m_0$ at $h = 0$. However, in actual experiments, one often sees hysteresis. The magnetization does not

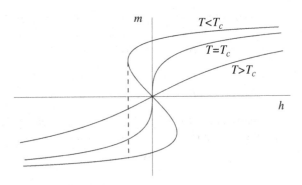

Fig. 5.2 The mean-field results of magnetization m v.s. magnetic field h for fixed temperatures, i.e., the isotherms.

jump to a negative value until the magnetic field is decreased to a sufficiently negative value, shown in the figure as a dotted line. The hysteresis behavior is not in thermal equilibrium but rather in meta-stable states.

5.2.1 Feynman-Jensen-Bogoliubov inequality

The mean-field equation, Eq. (5.17), gives us the equation of state, which is not a complete specification of the system. It would be better if we can give an expression for the free energy, which completely characterizes the system. For this, we can do a separate calculation based on $F = U - TS$. The Feynman, Jensen, and Bogoliubov variational method focuses on approximations to the free energy so that the equation of state is just a natural consequence of it.

To do this, we first introduce the Jensen inequality, which is a mathematical result: if $\phi(x)$ is a convex function, then $\overline{\phi(x)} \geq \phi(\overline{x})$. Here the overlines represent averages and x (which could be a vector) is a random variable. If x takes only two possible values, x_1 and x_2, we obtain the usual inequality for convex function

$$\lambda_1 \phi(x_1) + \lambda_2 \phi(x_2) \geq \phi(\lambda_1 x_1 + \lambda_2 x_2), \quad 0 \leq \lambda_1, \lambda_2 \leq 1, \ \lambda_1 + \lambda_2 = 1.$$
$$(5.25)$$

Feynman generalized this to the quantum case for the evaluation of free energy. Thus we shall use a quantum-mechanical notation of trace to denote the partition function, which can be replaced by a phase space integral for classical systems. We write the Hamiltonian as a sum of a reference system H_0 and a perturbation H'. The associated partition functions are

$$Z = \text{Tr}\left(e^{-\beta H}\right), \quad Z_0 = \text{Tr}\left(e^{-\beta H_0}\right), \quad H = H_0 + H'. \tag{5.26}$$

We consider the ratio of the partition functions,

$$\frac{Z}{Z_0} = \frac{\text{Tr}\left(e^{-\beta(H_0 + H')}\right)}{\text{Tr}\left(e^{-\beta H_0}\right)} = \left\langle e^{-\beta H'}\right\rangle_0 \geq e^{-\beta\langle H'\rangle_0}. \tag{5.27}$$

The angular brackets with a subscript 0 denote

$$\langle \cdots \rangle_0 = \frac{\text{Tr}\left(e^{-\beta H_0} \cdots\right)}{Z_0}. \tag{5.28}$$

Since $e^{\pm x}$ is a convex function with respect to x, in the last step in Eq. (5.27), we have used the Jensen inequality. Here we assume that the operators H_0 and H' commute so that $e^{-\beta(H_0 + H')} = e^{-\beta H_0} e^{-\beta H'}$.

However, as it turns out, the inequality ([Feynman (1972)], p. 67; [Callen (1985)], p. 434) does not depend on this assumption. Finally, taking the logarithm and dividing with a negative $(-\beta)$ which reverses the direction of the inequality, we obtain

$$F = -\frac{1}{\beta} \ln Z \leq F_0 + \langle H' \rangle_0. \qquad (5.29)$$

This is known as the Feynman-Jensen-Bogoliubov (FJB) or also referred to as Gibbs inequality.

Let us use the inequality to rederive the mean-field equation for the Ising model. We choose the unperturbed system as $H_0 = -h_{\text{eff}} \sum_{i=1}^{N} \sigma_i$, here h_{eff} is some effective magnetic field to be determined. Since this is an independent spin problem, its solution has been worked out from the study of paramagnets. The free energy associated with H_0 is given by

$$F_0 = -k_B T N \ln z_0 = -k_B T N \ln \left[2 \cosh(\beta h_{\text{eff}}) \right]. \qquad (5.30)$$

We compute the correction term to F_0 by

$$\langle H' \rangle_0 = \langle H - H_0 \rangle_0 = \left\langle -J \sum_{\langle i,j \rangle} \sigma_i \sigma_j - (h - h_{\text{eff}}) \sum_{i=1}^{N} \sigma_i \right\rangle_0$$

$$= -\frac{1}{2} J q N m^2 - (h - h_{\text{eff}}) m N. \qquad (5.31)$$

In obtaining the second line above, we have defined $m = \langle \sigma_i \rangle_0$, and since each spin is independent of the others, we have $\langle \sigma_i \sigma_j \rangle_0 = \langle \sigma_i \rangle_0 \langle \sigma_j \rangle_0 = m^2$, i.e., the spins are uncorrelated. The independent spin approximation is an important feature of the mean-field theory. The factor $1/2$ takes care of double counting. Each spin has q neighbors, so the total number of neighbors are qN, but we have counted each twice.

The FJB inequality says

$$F \leq -k_B T N \ln \left[2 \cosh(\beta h_{\text{eff}}) \right] - \frac{1}{2} J q N m^2 - (h - h_{\text{eff}}) m N \equiv \Psi. \qquad (5.32)$$

Here we have defined the right-hand side of the inequality as Ψ which depends on the parameters h_{eff} and m. The inequality is valid for any choices of h_{eff} and m. The value Ψ provides an upper bound to F. The best estimate to the free energy F is obtained if we minimize with respect to them, by

$$\frac{\partial \Psi}{\partial h_{\text{eff}}} = 0 \quad \rightarrow \quad m = \tanh(\beta h_{\text{eff}}), \qquad (5.33)$$

$$\frac{\partial \Psi}{\partial m} = 0 \quad \rightarrow \quad h_{\text{eff}} = h + Jqm. \qquad (5.34)$$

Combining the two results, we see that it is identical to the usual mean-field equation. The mean-field estimate to the Helmholtz free energy, after substituting the second equation, $h_{\text{eff}} = h + Jqm$, is

$$F \approx -k_B T N \ln\left[2 \cosh\left(\beta(h + Jqm)\right)\right] + \frac{1}{2} JqNm^2. \qquad (5.35)$$

Notice the sign change in the m^2 term due to the cancellation with the $(h - h_{\text{eff}})$ term. The value m still needs to be solved from the mean-field equation, as F is a function of temperature T and magnetic field h, and not m.

5.2.2 The cavity method

The critical temperature from the mean-field theory is given by $k_B T_c = qJ$. From the Table 5.1 we notice that this prediction is closer to the exact value when the dimension is higher. The claim is that the mean-field theory becomes exact in infinite dimensions. If we fix the coupling J and take $q = 2d$ to infinity, then T_c also runs to infinity. A better point of view is to set $J = J^*/q$, fixing J^* and taking $q \to \infty$. Then the critical value approaches a constant if the coupling is made small, such that $k_B T_c/J^* = 1$.

The statement that the mean-field theory becomes exact in infinite dimensions can be made rigorous by the cavity method [Georges *et al.* (1996)]. We take a d-dimensional hypercubic lattice and split the Hamiltonian into three terms:

$$H = -h\sigma_0 - J \sum_{j \in \text{ nn of } 0} \sigma_j \sigma_0 + H^0. \qquad (5.36)$$

Here we single out a particular site, and call it site 0. The first term is the single site in a magnetic field h. The second term is the couplings of the site 0 with the other sites belonging to the cavity. The last term is the cavity Hamiltonian. The cavity Hamiltonian is obtained if we remove the site 0 along with the associated couplings with it. We define the partition function of the cavity as

$$Z^0 = \sum_{\sigma_i, i \neq 0} e^{-\beta(H^0 - \sum_{j \neq 0} \eta_j \sigma_j)}. \qquad (5.37)$$

Here the spin of site 0 is excluded and we have introduced extra field η_j coupled to each spin in the cavity. The correlation function of the spins is

$$\langle \sigma_i \sigma_j \cdots \sigma_k \rangle_C^0 = \frac{\partial^n \ln Z^0}{\partial \beta \eta_i \, \partial \beta \eta_j \cdots \partial \beta \eta_k}\bigg|_{\eta_i = 0}. \qquad (5.38)$$

The subscript C denotes the "connected part", or "cumulant", which is defined by the right-hand side mixed derivatives with respect to $\ln Z^0$. We can use the cumulants to write a Taylor expansion of $\ln Z^0$ with respect to $\beta \eta_i$,

$$\ln Z^0(\eta) = \ln Z^0(\eta_i = 0) + \beta \sum_i \langle \sigma_i \rangle^0 \eta_i + \frac{1}{2} \beta^2 \sum_{ij} \langle \sigma_i \sigma_j \rangle^0_C \eta_i \eta_j + \cdots .$$

$$(5.39)$$

There is no distinction between $\langle \sigma_i \rangle^0$ and $\langle \sigma_i \rangle^0_C$ so we omit the subscript for this term. An important point is that all the second order or higher cumulants go to 0 as the dimension d goes to infinity. We can see this by a high-temperature expansion of the correlation, which is proportional to J. As a result, only the first two terms in the Taylor expansion survive in the infinite dimensional limit.

We define an effective single spin 0 Hamiltonian by

$$e^{-\beta H_{\text{eff}}(\sigma_0)} = \sum_{\sigma_i, i \neq 0} e^{-\beta H} = e^{\beta h \sigma_0} \sum_{\sigma_i, i \neq 0} e^{-\beta(H^0 - \sum_j \eta_j \sigma_j)}. \qquad (5.40)$$

The effective Hamiltonian is obtained by summing over all spins except the site 0. It is clear that this effective Hamiltonian reproduces all the properties that involve a single spin, for example the magnetization $\langle \sigma_0 \rangle_H$ with respect to the original Hamiltonian. In the last term we write in terms of a general cavity model, but to reproduce the nearest neighbor couplings between site 0 and the cavity, we must take $\eta_j = J\sigma_0$ if j belongs to the nearest neighbors of σ_0 and 0 otherwise. Taking the logarithm of the above equation and then dividing by $-\beta$, we obtain

$$H_{\text{eff}}(\sigma_0) = -h\sigma_0 - \sum_{j \in \text{ nn of } 0} \langle \sigma_j \rangle^0 \eta_j + \cdots . \qquad (5.41)$$

The difference between $\langle \sigma_j \rangle^0$ and $\langle \sigma_j \rangle$ is that $\langle \sigma_j \rangle^0$ is with respect to the cavity Hamiltonian with the site 0 (and couplings to it) removed, while $\langle \sigma_j \rangle$ is the exact magnetization with respect to the original full Hamiltonian H. If the dimension is high and the number of neighbors is large, removing one neighbor to the site j would make only a small effect, .i.e.,

$$\langle \sigma_j \rangle^0 = \langle \sigma_j \rangle + O\left(\frac{1}{q}\right). \qquad (5.42)$$

Using this result, taking the limit $q = 2d \to \infty$, and the fact that all sites are equivalent $\langle \sigma_j \rangle = m$, we find

$$H_{\text{eff}}(\sigma_0) = -h\sigma_0 - Jqm\sigma_0, \qquad (5.43)$$

which is the mean-field Hamiltonian. We have thus demonstrated that the mean-field theory is an exact theory in the limit of infinitely large dimensions.

In the mean-field treatment we have focused on a single site, ignoring the correlations, for example, we assumed $\langle \sigma_i \sigma_j \rangle \approx \langle \sigma_i \rangle \langle \sigma_j \rangle$, $i \neq j$. To improve the accuracy of prediction of the critical temperature T_c, we can take into account the nearest neighbor correlation by focusing on a larger cluster instead of a single site. At the lowest order of such a treatment, we can treat the site 0 and its q neighbors exactly (i.e., have an effective Hamiltonian for the $q + 1$ sites), and approximate the rest of the spins by mean fields. This leads to the Bethe approximation [Plischke and Bergersen (2006)], with much improved estimate for the critical temperature. However, the critical exponents remain the mean-field values no matter how large the cluster is. For example, the order parameter exponent in $m \sim a|T - T_c|^{\beta}$, is $\beta = 1/2$. By systematically analyzing how the amplitudes change with the cluster sizes (more precisely, how it changes with respect to the difference of mean-field predicted critical temperature and true thermodynamic limit critical temperature), one can estimate the true critical exponents beyond the mean-field prediction. For example, for the 2D Ising model, the exact value is $1/8$ instead of the mean-field $1/2$. This more advanced method is developed by Suzuki and is called the coherent anomaly method [Suzuki (1995)].

5.3 Landau theory

We have worked on the mean-field theory from three perspectives: the single site approximation, reducing a many-body problem to a single body problem, the Feynman-Jensen-Bogoliubov inequality relating the original problem with a simpler problem, and the cavity method which justifies the validity of mean-field result if dimensions are high. All of these are based on a concrete model. In the Landau theory, we argue at the level of thermodynamics, assuming certain plausible analytic behaviors. Taking the magnetic systems as an example, the "Helmholtz" free energy is $F = -\frac{1}{\beta} \ln Z$, and $Z = \sum_{\{\sigma\}} e^{-\beta H}$ is the partition function. Thus, the differential of the Helmholtz free energy is $dF = -SdT - Mdh$, and $M = Nm$ is the total magnetization.

The Landau theory works with the "Gibbs" free energy $G = F + Mh$ as a function of temperature T and order parameter M. The differential is

$$dG = -SdT + hdM. \tag{5.44}$$

We will make a few assumptions about this function G. (1) G is an even function of M. We can see this from the mean-field result. If we flip the sign of the field h, M also changes sign. If we eliminate h in favor of the order parameter M, it will be an even function of M. (2) G is analytic in M. Near the critical point, the free energy G has a Taylor expansion in M, so

$$G(T, M) = G(T, 0) + AM^2 + BM^4 + \cdots . \tag{5.45}$$

We have used assumption (1) so that the odd powers are absent. The last assumption (3) is to assume $A = A(T) = (T - T_c)a$, $a > 0$ and treat B as a constant. These three assumptions lead to a critical behavior consistent with mean-field theory and also consistent with the prediction of van der Waals theory for fluids. They all give the so-called mean-field critical exponents.

The thermodynamic properties of a magnetic system is completely contained in the function G. Figure 5.3 gives a qualitative picture of G vs. M for high and low temperature T. At the high temperature, $A > 0$, there is only one minimum value for M at 0. Below the critical temperature, $A < 0$, we have two minima and one maximum. The maximum represents an unstable equilibrium. The slope of the curve gives the equation of state for the magnetic system,

$$h = \frac{\partial G}{\partial M} = 2AM + 4BM^3 . \tag{5.46}$$

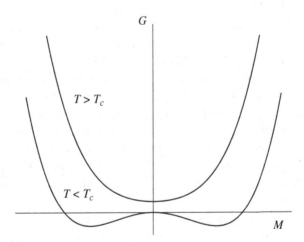

Fig. 5.3   The Gibbs free energy G vs. magnetization M in the Landau theory.

At $h = 0$, we have $(A + 2BM^2)M = 0$. This gives, either $M = 0$, or if $A < 0$, $M = \pm\sqrt{\frac{-A}{2B}}$. Since $A = (T - T_c)a$, we have the behavior of the order parameter

$$M \propto (T_c - T)^{1/2}, \quad T < T_c. \tag{5.47}$$

That is, $\beta = 1/2$. Exactly at T_c, $A = 0$, so the critical isotherm is $h = 4BM^3$, or $\delta = 3$. Finally, the susceptibility can be computed from

$$\chi = \frac{\partial M}{\partial h}\bigg|_{h=0} = \left(\frac{\partial h}{\partial M}\right)^{-1} = \left(\frac{\partial^2 G}{\partial M^2}\right)^{-1} = \frac{1}{2A + 12BM^2}. \tag{5.48}$$

The susceptibility exponent is $\gamma = 1$, but the coefficient is different above and below T_c. For $T > T_c$, $\chi = [2a(T - T_c)]^{-1}$, and for $T < T_c$, we have $\chi = [4a(T_c - T)]^{-1}$.

5.4   Ginzburg-Landau theory of superconductivity

Landau theory has been the paradigm of phase transitions established in the 1940s. This theory is phenomenological and is based on the concept of the existence of order parameters. In the 1970s, it was realized that the exponents predicted by Landau theory are not accurate. This gave birth to the renormalization group theory for critical phenomena. But the Landau theory based on a concept of the order parameter is still a very good guide to classifying phase transitions. Such concepts were central until recent years, when the concept of topological order and topological phases emerged, such as topological insulators. Before ending this chapter on mean-field theory based on Landau's concept, we discuss a generalization where the order parameter can be spatially dependent. We will think of a complex field Ψ which is the order parameter of a superconductor satisfying a Schrödinger-like equation.

The free energy density associated with the superconducting state of a superconductor is postulated to be [de Gennes (1966)]

$$f = f_n + \alpha|\Psi|^2 + \frac{\beta}{2}|\Psi|^4 + \frac{1}{2m}\left|(-i\hbar\nabla - 2e\mathbf{A})\,\Psi\right|^2 + \frac{1}{2\mu_0}\mathbf{B}^2. \tag{5.49}$$

Here $f = f(\mathbf{r})$ is the free energy density. The total free energy in some domain of space is the volume integral $F = \int f d^3\mathbf{r}$. In the above, the first three terms are very much like the usual terms in the Landau theory. The gradient term is the kinetic energy. This term takes into account spatial variation of the order parameter. The last term takes into account the magnetic field contribution. Here $\mathbf{A}$ is the vector potential, and $\mathbf{B} = \nabla \times \mathbf{A}$ is the magnetic induction. f_n is the free energy of the normal conductor, m is mass and $e < 0$ is electron charge. The factor of 2 accounts for the Cooper pairing of electrons. μ_0 is the vacuum permeability.

The actual physically realizable configuration of the order parameter Ψ and the vector field $\mathbf{A}$ is obtained by minimizing the total free energy. This is obtained by small variations on Ψ and $\mathbf{A}$. The minimum is obtained when $\delta F = 0$, giving a pair of equations,

$$\alpha \Psi + \beta |\Psi|^2 \Psi + \frac{1}{2m} \left(-i\hbar \nabla - 2e\mathbf{A} \right)^2 \Psi = 0, \qquad (5.50)$$

and

$$\mathbf{j} = \frac{1}{\mu_0} \nabla \times \mathbf{B} = \frac{e\hbar}{im} \left(\Psi^* \nabla \Psi - \Psi \nabla \Psi^* \right) - \frac{4e^2}{m} |\Psi|^2 \mathbf{A}. \qquad (5.51)$$

The first equation determines the order parameter Ψ, and the second equation gives current, which consists of a paramagnetic term (kinetic energy term) and a diamagnetic term. If the system is homogeneous, we have only the last term, which is essentially the London equation. Just like in the usual Landau theory, we can assume the parameter α is a linear function of temperature near T_c and make the assumption $\alpha \propto (T_c - T)$. In a bulk superconductor in equilibrium, $\mathbf{j} = 0$, $\mathbf{B} = 0$, and we can take $\mathbf{A} = 0$, so the order parameter takes the form $|\Psi| \propto (T_c - T)^{1/2}$ for temperature below T_c, which takes the usual mean-field value of $1/2$ for the order parameter critical exponent. Above T_c, $\Psi = 0$ by definition. The more elaborate BCS theory confirms this. As such, the conventional superconductor is well-described by a mean-field theory.

Problems

Problem 5.1. *The definition of heat capacity depends on the process, thus for a magnetic system like the Ising model, we can define two heat capacities, i.e., fixing magnetic field h, or fixing magnetization M. That is, $C_h = \left(\delta Q / dT \right)_h$, or $C_M = \left(\delta Q / dT \right)_M$. (a) Express the infinitesimal heat transfer δQ, in terms of appropriate thermodynamic functions for a magnetic system, thus give alternative expressions for the two heat capacities. (b) Based on the general thermodynamic relations, show that* *the difference is*

$$C_h - C_M = T \left(\frac{\partial M}{\partial T}\right)_h^2 \frac{1}{\chi_T},$$

where $\chi_T = (\partial M/\partial h)_T$ is the magnetic susceptibility.

Problem 5.2. *For an Ising paramagnetic model with the energy*

$$H(\sigma) = -h \sum_{i=1}^{N} \sigma_i,$$

show that the entropy can be expressed as

$$S = -N k_B \left[\frac{1+m}{2} \ln \frac{1+m}{2} + \frac{1-m}{2} \ln \frac{1-m}{2}\right],$$

where $m = \langle \sigma_i \rangle$.

Problem 5.3. *Consider the problem of a mean-field theory of an infinitely long one-dimensional Ising model, with the classical Hamiltonian (energy) given by $H = -J\sum_i \sigma_i \sigma_{i+1} - h\sum_i \sigma_i$, $\sigma_i = \pm 1$. Let us have the spins of site 1 and 2 treated explicitly and exactly and the rest of the spins of sites $\ldots, -2, -1, 0,$ and $3, 4, 5, \ldots,$ replaced by average, $m = \langle \sigma_i \rangle$. (a) Give the effective Hamiltonian for the two-spin system and derive the mean-field equations for the averages of two spins $\langle \sigma_1 \rangle$ and $\langle \sigma_2 \rangle$. (b) If $\langle \sigma_1 \rangle = \langle \sigma_2 \rangle = m$, discuss if there is a phase transition at $h = 0$, in the sense that $m > 0$ for $T < T_c$ and $m = 0$ for $T > T_c$ with $T_c > 0$. (c) Redo this problem by considering three sites, $0, 1,$ and $2,$ explicitly, and treating the rest by mean fields. Will the answer change?*

Problem 5.4. *Consider the standard ferromagnetic Ising model with nearest neighbor interactions in a magnetic field,*

$$H(\sigma) = -J \sum_{\langle i,j \rangle} \sigma_i \sigma_j - h \sum_{i=1}^{N} \sigma_i,$$

where each nearest neighbor interaction with coupling constant J is summed once only. We derive the mean-field equation in the following way. (a) First, split the Hamiltonian into two terms of the form, $H(\sigma) = H_{\text{cavity}} - \sigma_i h_i$, where H_{cavity} is the cavity Hamiltonian, and h_i depends on the spins of nearest neighbors of site i only. Give the explicit forms of H_{cavity} and h_i.

This will be helpful for the next step. (b) Prove an exact identity, known as the Callen identity:

$$\langle \sigma_i \rangle = \left\langle \tanh \left[\beta \left(h + J \sum_{j \,\in \text{nn of } i} \sigma_j \right) \right] \right\rangle,$$

where $\beta = 1/(k_B T)$. *The average has the usual meaning of* $\langle \cdots \rangle = \sum_\sigma \cdots e^{-\beta H}/Z$, *and the summation is over the nearest neighbor sites* j *of a fixed center site* i. *(c) Assuming that the spins are uncorrelated, in the sense,* $\langle \sigma_i \sigma_j \cdots \sigma_k \rangle \approx \langle \sigma_i \rangle \langle \sigma_j \rangle \cdots \langle \sigma_k \rangle$, *for any number of spins and for any sites* $i, j, \ldots$, *show that the usual mean-field equation is recovered.*

Problem 5.5. *One way to derive the mean-field theory more rigorously is to use Jensen's inequality,* $\overline{\phi(x)} \geq \phi(\overline{x})$. *(a) State the condition needed for the validity of the Jensen's inequality. (b) Based on the Jensen inequality or quantum version of the Feynman-Jensen (or Bogoliubov) inequality:* $F \leq F_0 + \langle (H - H_0) \rangle_0$, *derive the mean-field estimation of the free energy (the right-hand side) for a simplified Heisenberg model*

$$H = -J \sum_{\langle i,j \rangle} S_i^z S_j^z - h \sum_{i=1}^{N} S_i^z,$$

here $S_i^z = -S, -S+1, \ldots, S-1, S$, S *is either an integer or half integer.* i *or* j *denotes lattice site on a hyper-cubic lattice in* D *dimensions and the first sum is over the nearest neighbors only. (c) Derive the mean-field equation for* $\langle S^z \rangle$.

Problem 5.6. *For the Landau theory with the Gibbs free energy* $G(T, M) = G_0 + AM^2 + BM^4$, $A = a(T - T_c)$, G_0 *and* B *constant, determine the heat capacity* $C = \delta Q/dT$ *as a function of temperature* T *at zero magnetic field,* $h = 0$.

Problem 5.7. *Consider a Landau theory with an assumption of the Gibbs free energy as*

$$G(T, M) = G(T, 0) + a(T - T^*)M^2 - \frac{1}{3}cM^3 + \frac{1}{4}dM^4,$$

where a, c, *and* d *are some positive constants. (a) Show that the order parameter* M *is discontinuous at some transition point* T_c. *Determine* T_c. *Draw a qualitatively correct curve of* M *vs. temperature* T. *Pay attention to the stability of the solutions. The stable solution should be a global minimum of G with respect to M. (b) Find the latent heat Q of the first-order phase transition.*

Problem 5.8. *Starting from the Ginzburg-Landau free energy for a superconductor, Eq. (5.49), applying the variational principle, determine the pair of equations for the order parameter Ψ and the current $\mathbf{j}$.*

Chapter 6

Ising Models: Exact Methods

In the last chapter, we discussed mainly the mean-field theory as an approximation method to solve the Ising model and other interacting systems. In this chapter, we focus on a few available methods for exact treatment of the problems, but only for very specific systems amenable to exact solutions.

6.1 Transfer matrix method

The idea of the transfer matrix method is to write the computation of the partition function as a linear algebra problem of computing the trace of a matrix power, $Z = \mathrm{Tr}(P^N)$. The trace of a matrix is invariant under similarity transform, $P \to SPS^{-1}$:

$$\mathrm{Tr}(P^N) = \mathrm{Tr}(PP \cdots P) = \mathrm{Tr}(SPS^{-1}SPS^{-1}S \cdots SPS^{-1}). \qquad (6.1)$$

By a proper choice of a nonsingular matrix S, we can put SPS^{-1} into a Jordan form consisting of block-diagonal matrices with eigenvalues on the diagonal and 1 on the upper diagonal. Multiplications of upper triangular matrices are still upper triangular with eigenvalues λ_i^N on the diagonals. Thus, we find

$$Z = \sum_i \lambda_i^N, \qquad (6.2)$$

where λ_i is the eigenvalues of P determined from $\det(P - \lambda I) = 0$. The number of eigenvalues is the same as the dimension of the square matrix P. We can choose P such that it is real and symmetric. Then the eigenvalues are real. In the thermodynamic limit, i.e., $N \to \infty$, the free energy is dominated by the largest eigenvalue, then, $F = -k_B T N \ln \lambda_{\max}$.

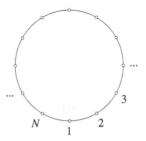

Fig. 6.1 The one-dimensional Ising model with a periodic boundary condition is equivalent to an Ising model defined on a ring.

Let us apply this technique to the one-dimensional Ising model with a Hamiltonian

$$H(\sigma) = -J\sum_{i=1}^{N}\sigma_i\sigma_{i+1} - h\sum_{i=1}^{N}\sigma_i = \sum_{i=1}^{N}E(\sigma_i,\sigma_{i+1}). \tag{6.3}$$

Here J is the nearest neighbor coupling, h is magnetic field, and spins take values $\sigma_i = \pm 1$. For the convenience of writing the summation over spins as a trace, we use periodic boundary conditions; see Fig. 6.1. This means that the site $N+1$ is the same as site 1, $\sigma_{N+1} = \sigma_1$. We can think of the Ising spins as sitting on a ring rather than a 1D line. We define the local energy as

$$E(\sigma_i,\sigma_{i+1}) = -J\sigma_i\sigma_{i+1} - h\sigma_i. \tag{6.4}$$

Then the partition function can be written as

$$\begin{aligned} Z &= \sum_{\{\sigma_i\}} e^{-\beta H(\sigma)} \\ &= \sum_{\sigma_1}\sum_{\sigma_2}\cdots\sum_{\sigma_N} e^{-\beta E(\sigma_1,\sigma_2)}e^{-\beta E(\sigma_2,\sigma_3)}\cdots e^{-\beta E(\sigma_N,\sigma_1)}. \end{aligned} \tag{6.5}$$

Note that the first and last index in the factors are the same σ_1. The other summations over σ_i take the form of matrix multiplication, if we define a matrix

$$P_{\sigma_i,\sigma_{i+1}} = e^{-\beta E(\sigma_i,\sigma_{i+1})} \tag{6.6}$$

or more explicitly a 2×2 matrix (indexed with $+1$ and -1, in that order)

$$P = \begin{pmatrix} e^{\beta(J+h)} & e^{\beta(-J+h)} \\ e^{\beta(-J-h)} & e^{\beta(J-h)} \end{pmatrix}. \tag{6.7}$$

Thus we have succeeded in identifying a matrix P such that $Z = \mathrm{Tr}(P^N)$.

The eigenvalues of matrix P can be found by solving the polynomial equation,

$$\det(P - \lambda I) = \det \begin{pmatrix} e^{\beta(J+h)} - \lambda & e^{\beta(-J+h)} \\ e^{\beta(-J-h)} & e^{\beta(J-h)} - \lambda \end{pmatrix} = 0. \qquad (6.8)$$

Here I is the identity matrix. λ satisfies a quadratic equation. The solutions can be easily found to be

$$\lambda_\pm = e^{\beta J} \cosh(\beta h) \pm \sqrt{e^{2\beta J} \sinh^2(\beta h) + e^{-2\beta J}}. \qquad (6.9)$$

It is clear that $\lambda_+ > \lambda_-$, so the free energy is $F = -k_B T \ln\left(\lambda_+^N + \lambda_-^N\right) \approx -N k_B T \ln \lambda_+$, in the thermodynamic limit. The total magnetization is obtained by taking the derivative with respect to h. After some algebra, we obtain,

$$M = -\frac{\partial F}{\partial h} = \frac{N}{\beta} \frac{\partial \ln \lambda_+}{\partial h} = N \frac{\sinh(\beta h)}{\sqrt{\sinh^2(\beta h) + e^{-4\beta J}}}. \qquad (6.10)$$

If we set $J = 0$, we get, using $\cosh^2 x - \sinh^2 x = 1$, $M/N = \tanh(\beta h)$. This is the same result as for a paramagnet. If $T \neq 0$, then $M \to 0$ as $h \to 0$, i.e., for a nonzero temperature, the system cannot support spontaneous magnetization. The system is always in the paramagnetic phase. There is no finite temperature phase transition in the 1D Ising model. When $T = 0$, we see that $M/N = +1$ or -1, depending on the sign of h. These are the ground states of the Ising model.

6.2 High-temperature expansion and pair correlation function

Alternative to the transfer matrix method is a high-temperature expansion. For simplicity, we now consider the case of zero magnetic field. The Hamiltonian only has the nearest neighbor couplings. Focusing on a particular bond connecting site i with $i + 1$, we have

$$e^{\beta J \sigma_i \sigma_{i+1}} = \cosh(\beta J) + \sigma_i \sigma_{i+1} \sinh(\beta J). \qquad (6.11)$$

This identity can be easily checked since the quantity $\sigma_i \sigma_{i+1} = \pm 1$ only. We introduce the short-hand notation $x = \tanh(\beta J)$, then the partition function is

$$Z = \sum_{\sigma_1, \sigma_2, \dots, \sigma_N} \prod_{i=1}^{N} \cosh(\beta J)\left(1 + \sigma_i \sigma_{i+1} x\right). \qquad (6.12)$$

If we expand the multiplicative factors $(1 + \sigma_1\sigma_2 x)(1 + \sigma_2\sigma_3 x) \cdots (1 + \sigma_N\sigma_1 x)$, we get various powers of x and the products of the spin variables. If we then sum over σ_i for all i, the odd power of σ_i adds up to 0, the survival terms are these with an even power in the spins. This can happen only in two ways for the 1D Ising model, the term of 1, and the term x^N where each spin appears exactly twice. As a result, we obtain an exact partition function for the 1D Ising model (without the magnetic field),

$$Z = 2^N \cosh^N(\beta J)\left(1 + x^N\right). \tag{6.13}$$

One can show that this agrees with the earlier result of the transfer matrix if we set $h = 0$.

The pair correlation function is defined as

$$g(j) = \langle \sigma_0\sigma_j \rangle = \langle \sigma_i\sigma_{i+j} \rangle = \sum_{\{\sigma\}} \sigma_0\sigma_j \frac{e^{-\beta H(\sigma)}}{Z}. \tag{6.14}$$

Imaging that the spins are on a ring, so site 0 is identified with site N. Due to the translational invariance of the system, the correlation function depends only on the distance j between the two spins. In particular, since $\sigma_i^2 = 1$, we have $g(0) = 1$.

To compute the pair correlation function, we perform a high-temperature expansion for the numerator and denominator. The denominator is the partition function Z whose expression is already known. The numerator has the extra feature of spins σ_0 and σ_j. In order to have an even number of spins, the only way is to have connected bonds (strings) with the x factors from site 0 to site j, which can be done in two ways (due to the periodic boundary condition, or ring geometry). Such bonds have length j or $N - j$. The $2^N \cosh^N(\beta J)$ factors cancel between the numerator and denominator. Thus, we have

$$g(j) = \frac{x^j + x^{N-j}}{1 + x^N}, \quad j = 0, 1, \ldots, N-1. \tag{6.15}$$

In the thermodynamic limit, we can drop the x^N term as $|x| < 1$, and write $g(j) \approx e^{-j/\xi}$, with $\xi = -1/\ln\tanh(\beta J) \approx \frac{1}{2}e^{2\beta J}$. ξ is called the correlation length. The correlation length diverges to infinity as temperature approaches 0 for the 1D Ising model. Roughly speaking, the correlation length measures how long the spin up remains up, before it flips in a typical thermal state. At the ground state, all spins are plus (or minus), the orientation of the spins correlates infinitely far.

It turns out that the correlation length ξ can also be obtained in a transfer matrix method, by the ratio of the largest eigenvalue to the next largest large eigenvalue, see [Yeomans (1992)], Chap. 5.

6.3 Duality relation

We study the Ising model defined on a planar graph. Examples of such graphs are shown in the figure. On each site, we associate with an Ising spin, σ_i, and the bond (a link) is associated with a ferromagnetic coupling J, with an interaction term $-J\sigma_i\sigma_j$ connecting the site i with j. We will not apply a magnetic field to it. A planar graph is either an infinitely large 2D lattice such as the square or triangular lattice, or it can be embedded onto a sphere without bond crossing. For each graph, we can construct a dual graph (or dual lattice) obtained in the following way: associate with each face of the original lattice a dual site (also called vertex). A face of the original graph consists of a plaquette (a closed polygon), bounded by the sites and links (edges). Connect the dual sites by dual link which always crosses the original link. This forms the dual graph. For example, the dual lattice of a square lattice is still a square lattice, shifted by half lattice spacings in both directions, see Fig. 6.2(b). We say that the square lattice is self-dual. The dual lattice of a triangular lattice turns out to be a honeycomb lattice. According to our definition, it is possible that the dual sites can be connected by the bonds (links) more than once.

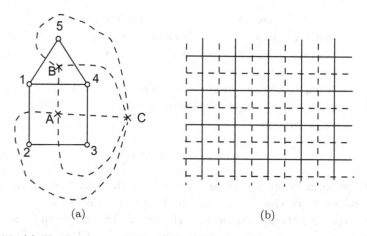

(a)   (b)

Fig. 6.2 (a) The 'house' graph (dot and solid lines) and its dual graph (crosses and dotted line). (b) Square lattice (solid lines) and its dual lattice (dotted lines).

There is a relation between the number of sites and links known as the Euler relation. Let N be the number of sites on the original lattice, and N^* the number of faces or plaquettes which is equal to the number of dual sites. Let $L = L^*$ be the number of links (same on the original lattice and dual lattice, since each link must be crossed by the dual link by construction). Then the Euler relation is

$$N - L + N^* = 2. \tag{6.16}$$

The number 2 is related to the fact that the embedding space is topologically equivalent to the surface of a sphere. If we embed the graph onto a torus (which is not planar), the number will be 0.

The duality relation is a relation about the partition function in a high-temperature expansion relating to a lower-temperature expansion on the dual lattice. We first define the expansions, and then demonstrate this relationship [Domb (1996)]. Using the identity Eq. (6.11), for the general planar graph, we have the partition function similar to Eq. (6.12) except that the product is over the links of the graph. If we multiply out the $1 + \sigma_i \sigma_j x$ factors and sum over the spins, only those that form loops give non-zero contribution to the partition function. We can collect the terms and write the result as a power series in x, as

$$Z(N, L, K) = 2^N (\cosh K)^L \left(\sum_{r=0}^{L} p(r) x^r \right), \quad x = \tanh(K). \tag{6.17}$$

Here, $K = \beta J$, $p(r)$ has a nice graph-theoretic interpretation: $p(r)$ is the number of distinct figures (graphs) with r lines composed of closed polygons (the polygons can be disconnected). The lines must meet an even number of times at a site. Here we define $p(0) = 1$.

As an example, we can consider the 'house' graph shown in Fig. 6.2(a). We have no loop, one loop on the square, the triangle, and the pentagon. So the partition function is

$$Z = 2^5 (\cosh K)^6 \left(1 + x^3 + x^4 + x^5\right). \tag{6.18}$$

We perform low-temperature expansion on the dual lattice. The low-temperature expansion is worked out if we compute the energy of the system, ordering them from low to high energy. The first term is the contribution from the ground state, which has $-\beta^* H = LK^*$. $L = L^*$ is the number of links, same on the original lattice and the dual lattice. The next term is higher in energy by flipping a spin. So in general we write

$$Z^* = \sum_{\{\sigma\}} e^{\beta^* J \sum_{\langle i,j \rangle} \sigma_i \sigma_j} = 2e^{K^* L^*} \left(\sum_{r=0}^{r_{\max}} v^*(r) e^{-2rK^*} \right). \tag{6.19}$$

Here again we define $v^*(0) = 1$. The factor of 2 is because of the degeneracy of the states; a state with spin σ_i and $-\sigma_i$ with all spins flipped has the same energy. The asterisk denotes quantities on the dual lattice. $v^*(r)$ gives the number of configurations with r + to − boundaries. For example, on a square lattice with all spins up at the ground state, if we flip one spin, we create + to − boundaries at four links, so $r = 4$. The duality says

"The number $p(r)$ for the high-temperature expansion is equal to the number $v^*(r)$ in a low-temperature expansion on the dual lattice."

To understand this relation, one can see that each high-temperature expansion graph can be mapped to a low-temperature graph where the link (the number of such is r) is mapped to a + to − boundary. On a finite planar graph, we should allow the possibility of multiple links on the dual lattice between two dual sites. For the 'house' graph, let us call the dual site A, B, and C, with C being the site outside the house. Then $N^* = 3$, $L = L^* = 6$, and the partition function on the dual lattice is

$$Z^* = \sum_{\sigma_A, \sigma_B, \sigma_C} e^{K^*(\sigma_A \sigma_B + 2\sigma_B \sigma_C + 3\sigma_A \sigma_C)}$$

$$= 2e^{6K^*} \left(1 + e^{-6K^*} + e^{-8K^*} + e^{-10K^*} \right). \tag{6.20}$$

The correspondence between the low-temperature and high-temperature expansions can be established at the configuration level.

Duality means that we do not need to do two separate calculations, one calculation automatically implies another calculation, so the partition functions for the two problems, one for high temperature on the original lattice and another low temperature at the dual lattice is related by

$$\frac{Z(N, L, K)}{2^N (\cosh K)^L} = \frac{Z^*(N^*, L^*, K^*)}{2 \, e^{K^* L^*}} \tag{6.21}$$

if we choose

$$x = \tanh K = e^{-2K^*}. \tag{6.22}$$

Then, something remarkable happens for graphs that are self-dual. For the infinitely large square lattice, the dual lattice is still a square lattice. This means our high-temperature expansion is also a low-temperature expansion after a variable transform. Thus, every point K is mapped to a new point K^* through the map, Eq. (6.22). Let us assume for the free energy $F = -k_B T \ln Z$, there is only one singular point with respect to temperature, or $K = J/(k_B T)$. Then the critical temperature T_c will be mapped to T_c^*. If $T_c \neq T_c^*$, then we could have two singular points, contradicting our assumption, thus, we expect, for the two-dimensional Ising model on a square lattice that, the critical value is determined by $T_c = T_c^*$, or

$$\tanh K_c = e^{-2K_c} \rightarrow K_c = \frac{J}{k_B T_c} = \frac{1}{2} \ln(\sqrt{2} + 1) \approx 0.44068679. \quad (6.23)$$

The mean-field value, $K_c = 1/q = 0.25$, is wrong by about a factor of 2. This remarkable result was first obtained by Kramers and Wannier in 1941 through the duality argument and was later confirmed by Onsager [Onsager (1944)] from his exact calculation of the partition function. With some twists, we can also work out T_c for the triangular lattice and honeycomb lattice, but duality does not hold in three dimensions.

6.4   Supplementary reading: Exact solution of 2D Ising model

The exact solution of the Ising model in two dimensions is a great victory of statistical mechanics, solved by Onsager in 1944. The mathematics for the solution is extremely involved. We first just state the results.

(1) The critical temperature T_c is determined by the equation $\sinh \frac{2J}{k_B T_c} = 1$, or equivalently $\tanh \frac{J}{k_B T_c} = \sqrt{2} - 1$, or $e^{-2J/k_B T_c} = \sqrt{2} - 1$. Solving one of these equations, the dimensionless temperature has the value $k_B T_c/J \approx 2.269$.

(2) The average energy per site (total energy divided by the number of spins) in zero magnetic field $(h = 0)$ is

$$u = -J \left\{ 1 \pm \frac{2}{\pi} \sqrt{1 - z^2} \, I(z) \right\} \coth 2K, \quad K = \frac{J}{k_B T}, \quad (6.24)$$

where the $+$ sign is for $T < T_c$ and $-$ sign is for $T > T_c$, and $I(z) = \int_0^{\pi/2} \frac{d\omega}{\sqrt{1 - z^2 \sin^2 \omega}}$, $z = \frac{2 \sinh 2K}{\cosh^2 2K}$.

(3) The derivative of u with respect to T is the specific heat. Near the critical point it is asymptotically

$$\frac{c}{k_B} \approx \frac{2}{\pi} \left(\ln \coth \frac{\pi}{8} \right)^2 \left[\ln \left(\frac{\sqrt{2}}{|K - K_c|} \right) - 1 - \frac{\pi}{4} \right]. \qquad (6.25)$$

(4) The spontaneous magnetization ($h \to 0^+$) per site is

$$m = \left(1 - \frac{1}{\sinh^4 2K} \right)^{1/8} \approx (T_c - T)^{1/8}, \quad T < T_c. \qquad (6.26)$$

This last result was first obtained by C. N. Yang. Of course, $m = 0$ for $T \geq T_c$.

We now briefly outline the solution methods. The original method of Onsager uses transfer matrices and is improved by Kaufmann, see Chap. 15 in [Huang (1987)]. This is based on the algebraic properties of gamma matrices, $\Gamma_i \Gamma_j + \Gamma_j \Gamma_i = 2\delta_{ij}$, two matrices for each row of spins of a 2D lattice. Alternatively, we can consider combinatorial methods by considering the high-temperature expansion of the partition function, as used in demonstrating the duality relation. This amounts to a correct counting of the closed polygons on square lattice. This can be done by a random walk with a phase factor, depending on the step taken — straight ahead is weighted with 1, left turn $e^{i\pi/4}$, right turn $e^{-i\pi/4}$, and 180-degree with weight 0, see [Feynman (1972)], Chap. 5.4. Another method [Isihara (1971)] is to write the partition function in terms of a Pfaffian, which is the square root of the determinant of an anti-symmetric matrix, $\sqrt{\det(A)}$, $A^T = -A$. The matrix A is then diagonalized, given an explicit expression for the partition function. Since the Pfaffian is closely related to Gaussian integral of Grassmann numbers, solution based on Grassmann algebra also appears. Here we follow [Plischke and Bergersen (2006)] which was originally due to Schultz *et al.* [Schultz *et al.* (1964)] by transforming the classical 2D Ising problem into a one-dimensional quantum fermion problem and then diagonalizing the matrices.

The first step is standard. Consider an $N \times N$ square lattice with periodic boundary conditions. We write the partition function in terms of transfer matrices,

$$Z = \text{Tr} \, (V_1 V_2)^N. \qquad (6.27)$$

The size of the matrices are $2^N \times 2^N$, which maps a column of spins into another column. It is convenient to represent the matrices by Pauli spins

Advanced Statistical Mechanics

σ_j^x and σ_j^z on each site $j = 1, 2, \ldots, N$.

$$V_1 = e^{K \sum_{j=1}^{N} \sigma_j^z \sigma_{j+1}^z}, \quad V_2 = (2 \sinh 2K)^{N/2} e^{K^* \sum_{j=1}^{N} \sigma_j^x}. \tag{6.28}$$

Here $K = J/(k_B T)$, $\tanh K^* = e^{-K}$, V_1 is diagonal and is the term due to interactions within a column and V_2 is the interaction between the neighboring columns. The Pauli spins at different sites should be viewed as forming a direct product. In particular, we can think of the transfer matrices V_1 and V_2 as operators in a Hilbert space of quantum spins spanned by the eigenstates of σ_j^z, such that $\sigma_j^z |\sigma_1 \sigma_2 \cdots \sigma_j \cdots \rangle = \sigma_j |\sigma_1 \sigma_2 \cdots \sigma_j \cdots \rangle$.

Our next step is to transform the spin variables $\sigma_j^{x,z}$ into fermion variables c_j and $c_j^\dagger$. This is accomplished by the Jordan-Wigner transform. We introduce the raising and lowering operators (in x direction) by $\sigma_j^x = 2\sigma_j^+ \sigma_j^- - 1$, and $-\sigma_j^z = \sigma_j^+ + \sigma_j^-$, then the transform is

$$(\sigma_j^-)^\dagger = \sigma_j^+ = e^{\pi i \sum_{m=1}^{j-1} c_m^\dagger c_m} c_j^\dagger. \tag{6.29}$$

Here the fermion operators satisfy $c_j c_k^\dagger + c_k^\dagger c_j = \delta_{jk}$ and $c_j c_k = -c_k c_j$. It can be checked that the original spin commutation relations are preserved. After transforming into fermion operators, the matrices become

$$V_1 = e^{K \sum_{j=1}^{N} (c_j^\dagger - c_j)(c_{j+1}^\dagger + c_{j+1})}, \tag{6.30}$$

$$V_2 = (2 \sinh 2K)^{N/2} e^{K^* \sum_{j=1}^{N} (2c_j^\dagger c_j - 1)}. \tag{6.31}$$

Here in obtaining the fermion operator representation, we note that all the occupation number operators $c_j^\dagger c_j$ commute with each other. This gives V_2 without trouble. For V_1, we must use periodic boundary $c_{N+1} = c_1$ if total fermion $n = \sum_j c_j^\dagger c_j$ is odd, and anti-periodic boundary if n is even. Due to the presence of cc and $c^\dagger c^\dagger$ terms, the situation is very much like the BCS theory where fermion number is not a conserved quantity.

The 3rd step is to go into Fourier space, by

$$c_j = \frac{1}{\sqrt{N}} \sum_q a_q e^{iqj}. \tag{6.32}$$

The transformation for the creation operators is obtained by hermitian conjugate. The inverse transform takes the same form with the replacement of i by $-i$. Due the requirement of period or anti-periodic boundary condition, we take $j = \pm 1, \pm 3, \ldots, \pm(N-1)$ for n even, and $j = 0, \pm 2, \pm 4, \ldots, \pm(N-2), N$ for n odd. Since the operators with different

q commute, we can write

$$V = V_2^{1/2} V_1 V_2^{1/2} = (2 \sinh 2K)^{N/2} \prod_{q>0} V_q, \quad V_q = V_{2q}^{1/2} V_{1q} V_{2q}^{1/2}, \qquad (6.33)$$

where

$$V_{1q} = \exp\left\{ 2K \cos(q)(a_q^\dagger a_q + a_{-q}^\dagger a_{-q}) - 2iK \sin(q)(a_q^\dagger a_{-q}^\dagger + a_q a_{-q}) \right\},$$

$$V_{2q} = \exp\left\{ 2K^* (a_q^\dagger a_q + a_{-q}^\dagger a_{-q} - 1) \right\}. \qquad (6.34)$$

The operator V_q for each $q > 0$ is spanned by a much smaller 4-dimensional space. We can take the matrix elements with vacuum state $|0\rangle$, one-particle states $|p\rangle = a_q^+|0\rangle$ and $|-p\rangle = a_{-p}^+|0\rangle$, and two-particle state $|2\rangle = a_p^+ a_{-p}^+|0\rangle$. It can be checked that the matrix V_q is already diagonal in the two-dimensional space spanned by $|p\rangle$ and $|-p\rangle$, with degenerate eigenvalue $e^{2K\cos q}$. As a result, we just need to focus on the other two-dimensional space spanned by $|0\rangle$ and $|2\rangle$. This 2×2 matrix can be diagonalized, giving the two eigenvalues

$$\lambda_\pm(q) = e^{2K \cos q \pm \epsilon(q)}, \quad \epsilon(q) \geq 0, \qquad (6.35)$$

$$\cosh \epsilon(q) = \cosh 2K \cosh 2K^* + \cos q \, \sinh 2K \sinh 2K^*. \qquad (6.36)$$

The last step is to identify the largest eigenvalues for each q, and multiplying them for the overall transfer matrix. Here I will omit some details and subtlety. The free energy per spin in the thermodynamic limit ($N \to \infty$) is given by

$$\frac{F}{N} = -\frac{1}{2\beta} \ln(2 \sinh 2K) - \frac{1}{4\pi\beta} \int_{-\pi}^{\pi} \epsilon(q) dq. \qquad (6.37)$$

Problems

Problem 6.1. *Consider the Potts model defined by the Hamiltonian*

$$H = -J \sum_{i=1}^{N} \delta_{\sigma_i \sigma_{i+1}}, \quad \sigma_i = 1, 2, \ldots, q,$$

where the state is specified by discrete integers from 1 to q. The energy of the nearest neighbor is $-J$ if the states are the same and 0 otherwise. Periodic boundary condition is assumed, that is, $\sigma_{N+1} = \sigma_1$. (a) Write down the transfer matrix P for the one-dimensional, q-state Potts model. (b) Find the largest eigenvalue of the transfer matrix P. (c) Write down the expression for the free energy and find the correlation length of the model.

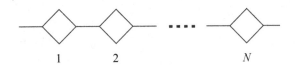

1   2   N

Fig. 6.3   Quasi-1D Ising chain of rhombuses.

Problem 6.2. *Consider a quasi-one-dimensional chain of Ising spins formed by connected rhombuses as shown above, Fig. 6.3. Each site has a spin without a magnetic field, with a ferromagnetic coupling constant J between the sites connected by a line. We assume the units are repeated exactly N times with the periodic boundary condition. Determine the partition function Z in two ways: (a) Use the transfer matrix method. (b) From high-temperature expansion, and show they are the same.*

Problem 6.3. *Certain class of 1D quantum problems can be mapped to 2D classical problems. One good example is the 1D quantum transverse field Ising model to 2D Ising model. Consider* $\hat{H} = -J \sum_i \hat{\sigma}_i^z \hat{\sigma}_{i+1}^z - g \sum_i \hat{\sigma}_i^x$, *here* $\hat{\sigma}_i^z$ *and* $\hat{\sigma}_i^x$ *are Pauli matrices of the z and x components at site i. Applying the Trotter-Suzuki formula,* $e^{\hat{A}+\hat{B}} = \lim_{M\to\infty} \left(e^{\hat{A}/M} e^{\hat{B}/M} \right)^M$, *we can write the quantum partition function* $\mathrm{Tr}\, e^{-\beta \hat{H}}$ *as product of matrices in the β direction. (a) Show that*

$$\langle \sigma_{i,k} | e^{a\hat{\sigma}_i^x} | \sigma_{i,k+1} \rangle = \sqrt{\frac{1}{2} \sinh(2a)}\, e^{-\frac{1}{2}\sigma_{i,k}\sigma_{i,k+1} \log \tanh(a)}.$$

Here $|\sigma_{i,k}\rangle$ *is the eigenvector of Pauli spin* $\hat{\sigma}_i^z$ *with eigenvalue* $\sigma_{i,k}$. *(b) From this result, show that the partition function of the 1D quantum transverse field Ising model is equivalent to the partition function of a 2D Ising model with nearest neighbor couplings* $K_1 = \beta J/M$, *and* $K_2 = \log \coth(\beta g/M)$, *i.e.,*

$$Z = \mathrm{Tr}\, e^{-\beta \hat{H}} = Z_0 \sum_{\sigma_{i,k}} e^{\sum_{i,k}(K_1 \sigma_{i,k}\sigma_{i+1,k} + K_2 \sigma_{i,k}\sigma_{i,k+1})}.$$

Here Z_0 *is some σ independent constant.*

Problem 6.4. *Consider an Ising model defined on the graphs shown below in Fig. 6.4, known as Cayley trees. The first three generations of the trees are shown. We assume that each site denoted by an open circle has an Ising spin* $\sigma_i = \pm 1$ *and each link has a nearest neighbor interaction,* $-J\sigma_i\sigma_j$.

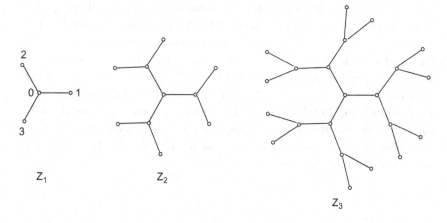

Fig. 6.4   Cayley trees.

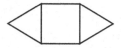

Fig. 6.5   Graph of one square with two triangles.

For example, the first generation of the graph is associated with the energy $E = -J\sigma_0\sigma_1 - J\sigma_0\sigma_2 - J\sigma_0\sigma_3$. *(a) Determine the canonical partition function* Z_1 *and* Z_2 *of the Ising model on the first and second generation Cayley trees. (b) Derive a general formula for* Z_N *for the* N-*th generation Cayley tree. (c) Discuss if the system has a phase transition at a finite temperature* $T_c > 0$ *when* N *approaches infinity.*

Problem 6.5.   *Consider the Ising spins on a finite lattice (or graph) of one square and two triangles as shown in Fig. 6.5. (a) Draw the dual lattice of the given lattice. Give the number of links* L, *number of sites* N, *and number of faces* N^* *(plaquettes) in both the original lattice and the dual lattice and show that Euler's relation is satisfied. (b) Draw the diagrams which have a nonzero contribution to the partition function* Z. *Give the high-temperature series expansion of* Z *in variable* $x = \tanh\big(J/(k_B T)\big)$. *(c) Use the duality relation to find the low-temperature expansion of the partition function* Z^* *on the dual lattice.*

Problem 6.6.   *The duality relation also exists for the two-dimensional Potts model. The q-state Potts model is similar to the Ising model except* *that the values of the spins σ_i take integers $1, 2, \ldots, q$. When $q = 2$ it is equivalent to the Ising model. The Hamiltonian is given by $H = -J \sum_{\langle ij \rangle} \delta_{\sigma_i \sigma_j}$. The energy of the nearest neighbor interaction is $-J$ if the values of spins at site i and j are the same, and 0 otherwise. (a) Consider a high-temperature expansion of the partition function Z on a square lattice with appropriate expansion variable x, give the first three nonzero terms of the result. (b) Develop a low-temperature expansion of Z for the same square lattice, starting from the ground state (which is q-fold degenerate), also obtain the first three nonzero terms. (c) Use the duality argument to determine the transition temperature T_c.*

Chapter 7

Critical Exponents, Scaling, and Renormalization Group

7.1 Critical exponents

Earlier we have introduced the concept of critical exponents. For example, the order parameter exponent is called β and heat capacity α. The exponent associated with the correlation length is called ν, and the correlation length itself is defined by the correlation function $g(r)$, which has an additional exponent η characterizing a power-law decay at critical point (when $\xi \to \infty$). The notation $m \sim |t|^{\beta}$ means, asymptotically when $t = (T - T_c)/T_c$ is small, the ratio $m/|t|^{\beta}$ approaches a constant when $t \to 0$. In Table 7.1, we list the values of the exponents for the two-dimensional Ising model, three-dimensional Ising model (which should have the same exponents as a fluid), as well as the mean-field predictions. We see that they are quite different and the mean-field results are not reliable.

7.2 Scaling relations

Are all the exponents independent? Is there a relation among the exponents? It turns out there are precise relations among them [Stanley (1971); Thompson (1988)]. Their relations can be explained (phenomenologically though) by Widom's scaling hypothesis for the free energy. Let us define the free energy per spin as $f(t,h) = F(T,h)/N$, here t is the fractional deviation from T_c and N is the number of spins. With this definition we postulate

$$f(t,h) = b^{-d}f(b^y t, b^x h) + \text{regular part of free energy.} \qquad (7.1)$$

Table 7.1 Definition and comparison of critical exponents.

definition	2D Ising	3D Ising/fluid	mean-field
$C \sim |t|^{-\alpha}$	0	0.11	?
$m \sim |t|^{\beta}$	1/8	0.326	1/2
$\chi \sim |t|^{-\gamma}$	7/4	1.24	1
$h \sim m^{\delta}$	15	4.79	3
$\xi \sim |t|^{-\nu}$	1	0.63	1/2
$g(r) \sim e^{-r/\xi}/r^{d-2+\eta}$	1/4	0.036	0

The first term on the right-hand side says that f is a generalized homogeneous function of its arguments, and the extra piece is "regular", meaning that it is smooth and analytic. For the rest of the discussion, we will just ignore it. Function f of such a form with arbitrary (but positive) b and with the regular part set to 0 is said to have a scaling form. There are only two independent exponents here, namely, y associated with t and x associated with the magnetic field h. Here d is the dimensionality of the system. We demonstrate that the two exponents x and y are sufficient to compute the other exponents, α, β, γ, and δ.

We first compute the order parameter exponent β associated with $m = -\partial f/\partial h$. For the convenience of computing the partial derivative with respect to h, we can set the first argument to 1 by the requirement $b^y t = 1$. This is possible because b is arbitrary; we take $b = t^{-1/y}$. Then the functional form of the free energy becomes

$$f(t, h) = t^{d/y} f(1, t^{-x/y} h) = t^{d/y} \hat{f}(t^{-x/y} h). \qquad (7.2)$$

Taking the partial derivative by applying the chain rule, we find

$$m = -\frac{\partial f}{\partial h} = -t^{\frac{d}{y}} \hat{f}'(t^{-\frac{x}{y}} h) t^{-\frac{x}{y}} \bigg|_{h \to 0} = -t^{\frac{d-x}{y}} \hat{f}'(0) \propto t^{\beta}. \qquad (7.3)$$

Here we made a tacit assumption that the function $\hat{f}$ goes to a constant when its argument goes to 0, for if that is not the case, then it should be factored in the t to some power of the prefactor. If this is indeed valid, then we can identify

$$\beta = \frac{d - x}{y}. \qquad (7.4)$$

The susceptibility exponent γ is obtained by taking the derivative again with respect to h, which results in the second derivative $\hat{f}''$, assuming it is a constant when $h \to 0$. Thus,

$$\gamma = \frac{2x - d}{y}. \tag{7.5}$$

For the δ exponent, we start from Eq. (7.3), and let $u = t^{-\frac{x}{y}} h$, we have $t = (u/h)^{-\frac{y}{x}}$, so

$$m = -t^{\frac{d-x}{y}} \hat{f}'(t^{-\frac{x}{y}} h) = -(u/h)^{-\frac{d-x}{x}} \hat{f}'(u) \Big|_{t \to 0} \propto h^{1/\delta}. \tag{7.6}$$

When we take the limit $t \to 0$, $u \to \infty$ (since x and y are positive numbers). We assume in that limit the function involving u approaches a constant, and we can identify

$$\delta = \frac{x}{d - x}. \tag{7.7}$$

The heat capacity exponent α can be computed by taking derivative with respect to t. Since $C_h = dQ/dT|_{h=0} = T dS/dT \approx T_c \partial S/\partial t|_{h=0}$, and $S = -N \partial f/\partial t$. The heat capacity per spin is $c = -T_c \partial^2 f/\partial t^2|_{h=0}$. Using a similar technique of taking the limit $h \to 0$ and noting the most singular part of the divergence, we find

$$c \propto t^{\frac{d}{y} - 2} \hat{f}(0) \propto t^{-\alpha} \quad \to \quad \alpha = 2 - \frac{d}{y}. \tag{7.8}$$

This last equation is known as hyperscaling (also known as Josephson's scaling law). Using the above equations relating α, β, γ, and δ to x and y, we can derive another set of relations. For example,

$$\alpha + 2\beta + \gamma = 2 - \frac{d}{y} + 2 \left(\frac{d - x}{y} \right) + \frac{2x - d}{y} = 2. \tag{7.9}$$

This is known as Rushbrook's scaling law. Analogously, we can verify Widom's scaling law,

$$\beta(\delta - 1) = \gamma. \tag{7.10}$$

There are two more relations that merely depend on the definitions of the correlation function. The integral of the correlation function gives the susceptibility. This gives the Fisher scaling law,

$$\gamma = \nu(2 - \eta), \tag{7.11}$$

and finally, the exponent $y = 1/\nu$.

In analyzing the numerical results, it is useful to consider a finite system of linear dimension L. In such a case, we also have the finite-size scaling for the free energy

$$f(t, h, L) = b^{-d} f(b^y t, b^x h, b/L). \tag{7.12}$$

This increases the arguments of the free energy to three variables. In the scaling hypothesis above this effectively reduces the arguments from three to two. Although the scaling hypothesis looks mysterious, it is fully justified by the renormalization group theory.

7.3　Renormalization group methods

The idea of renormalization group [Ma (1976); Amit and Martin-Mayor (2005)] is to consider the set of all Hamiltonians of a given problem and ask what happens to the Hamiltonians after we make some transformations to the dynamic variables. The variables are in some sense grouped or coarsened so that they describe the problem at a longer length scale. This can be done in momentum space, as has been done by K. Wilson using field-theoretic techniques. Alternatively, one can perform block-spin transformations, such that a group of, say, four spins, is treated as one single spin, using a majority rule.

A remarkable outcome of such a consideration is that one can focus directly on the critical exponent as a Taylor expansion in the dimensionality around 4, $\epsilon = 4 - d$. The dimension 4 is special as above 4 mean-field theory works. The order parameter exponent is given by

$$\beta = \frac{1}{2} - \frac{3\epsilon}{2(n+8)} + \frac{(n+2)(2n+1)\epsilon^2}{2(n+8)^3} + \cdots . \tag{7.13}$$

Up to ϵ^5 is known for this series presently [Vasil'ev (2004)]. Here n is the dimension of the spin, $n = 1$ for the Ising spin, $n = 2$ for the XY model, and $n = 3$ for the Heisenberg 3D spin.

Most of the calculations cannot be done easily, particularly the field-theoretic approach. Here, for the purpose of illustration of the basic idea [Huang (1987)], we consider a real space approach by spin decimation. The spin decimation removes, or rather, integrates out half of the spins, leaving a new system with a small number of degrees of freedom. We consider the relationship of the effective Hamiltonians between the original and new system, keeping the partition functions invariant. This operation can be done exactly in one dimension.

Thus, we consider a 1D Ising model [Yeomans (1992)]

$$\beta H = \bar{H} = -K \sum_{i=1}^{N} \sigma_i \sigma_{i+1} - h \sum_{i=1}^{N} \sigma_i - \sum_{i=1}^{N} C. \tag{7.14}$$

Here we absorb the inverse temperature into the definition of the Hamiltonian such that the partition function is given by $Z = \sum_{\{\sigma\}} e^{-\bar{H}}$. The 1D

Ising model is characterized by three model parameters, K, h, and C. The constant C is necessary since if we start with a model of $C = 0$, it will be generated by the transformation.

We keep the odd numbered sites untouched and sum over the even numbered sites. That is, consider the quantity

$$\sum_{\sigma_2, \sigma_4, \ldots} e^{-\bar{H}}. \tag{7.15}$$

Clearly, the result is a function of the remaining spin variables $\sigma_1, \sigma_3, \ldots$. We can view this result as the effect of some transformed Hamiltonian $\bar{H}'(\sigma_1, \sigma_3, \ldots)$ such that $e^{-\bar{H}'}$ is equal to it. Clearly, now $Z' = Z$. Here Z is the partition function of the original Hamiltonian with N spins and Z' is the partition function of the transformed Hamiltonian of $N/2$ spins. We can assume N even. Our mathematical task is to obtain the explicit form of the transformed Hamiltonian $\bar{H}'$. In general the transformed Hamiltonian could be very complicated and it may require many more parameters. Luckily for the 1D Ising model, the same three parameters are enough, but with new values K', h', and C'. The transformation from the triplet (K, h, C) to (K', h', C') is the RG transform of the 1D Ising model.

Let us focus on the summation of the even-sited terms, we have

$$\sum_{\{\sigma_2, \sigma_4, \ldots\}} \prod_{i=2,4,\ldots} e^{K\sigma_i(\sigma_{i-1}+\sigma_{i+1})+h\sigma_i+h(\sigma_{i-1}+\sigma_{i+1})/2+2C}. \tag{7.16}$$

After the summation of even site spins, we are left with odd sited spins. $\sigma_i = \pm 1$, we get

$$\prod_{i=2,4,\ldots} \left(e^{K(\sigma_{i-1}+\sigma_{i+1})+\frac{h}{2}(2+\sigma_{i-1}+\sigma_{i+1})+2C} \right. \tag{7.17}$$

$$\left. + e^{-K(\sigma_{i-1}+\sigma_{i+1})+\frac{h}{2}(-2+\sigma_{i-1}+\sigma_{i+1})+2C} \right). \tag{7.18}$$

The idea here is to rewrite the factors in the product in a form similar to the original Hamiltonian terms, in a symmetric form, i.e., we demand that the above expression is equal to

$$\prod_{i=2,4,\ldots} e^{K'\sigma_{i-1}\sigma_{i+1}+h'(\sigma_{i-1}+\sigma_{i+1})/2+C'}. \tag{7.19}$$

This is the same Ising model in one dimension, with half of the number of spins, and with a set of different model parameters. There are four possible choices of the spin values, $\sigma_{i-1} = \pm 1$, and $\sigma_{i+1} = \pm 1$. This gives us three independent equations to determine the transformation:

$$e^{K'+h'+C'} = e^{2K+2h+2C} + e^{-2K+2C}, \tag{7.20}$$

$$e^{K'-h'+C'} = e^{2K-2h+2C} + e^{-2K+2C}, \tag{7.21}$$

$$e^{-K'+C'} = e^{h+2C} + e^{-h+2C}. \tag{7.22}$$

We introduce a new set of variables by $w = e^{-4C}$, $x = e^{-4K}$, and $y = e^{-2h}$. The equations can be solved by Eq. (7.21) divided by (7.20), to obtain

$$y' = \frac{(x+y)y}{1+xy}. \tag{7.23}$$

Squaring the last equation, (7.22), and dividing by the product of the first two equations, we get

$$x' = \frac{x(1+y)^2}{(x+y)(1+xy)}. \tag{7.24}$$

Lastly, if we square the last equation, and multiply by the product of the first two equations, we obtain

$$w' = \frac{w^2 x y^2}{(x+y)(1+xy)(1+y)^2}. \tag{7.25}$$

These three equations form the renormalization group transformation $(x, y, w) \to (x', y', w')$ from the original 1D Ising model with a lattice spacing a to $2a$. This implies a scaling factor of $b = 2$. Note that the last equation is decoupled from the first two of x and y. This last equation gives a background free energy change. The real important features are the (x, y) variables which describe the flow of the renormalization group transformations.

The fixed points of the transformation and the behavior around the fixed points contain useful information. The point $(x^*, y^*) = (0, 1)$ is an unstable fixed point which corresponds to zero temperature and zero magnetic field. In fact, this is the "critical point" of the 1D Ising model. It is unstable as any deviation will lead to a flow into the infinite high temperature fixed line at $x = 1$ ($K = 0$ thus $T \to \infty$), with the exception when temperature is strictly at $T = 0$ with a magnetic field $h \neq 0$. In this case it flows to the infinite magnetic field fixed point at $(x, y) = (0, 0)$. See Fig. 7.1.

We linearize the transformation around the unstable fixed point $(0, 1)$ and introduce the deviations, $x = x - x^*$, and $\epsilon = y - y^* = y - 1$, then the

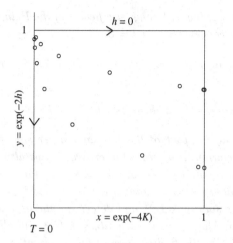

Fig. 7.1   The flow diagram for the RG iterations of 1D Ising model.

transformation simplifies to

$$x' \approx 4x = b^2 x, \quad \epsilon' \approx 2\epsilon = b\epsilon. \tag{7.26}$$

Since the renormalization group transformation preserves the partition functions, $Z' = Z$, this implies the relation between the free energy per spin of the old and transformed systems, that is

$$f_s(x, \epsilon) = b^{-1} f_s(x', \epsilon') = b^{-1} f_s(b^2 x, b\epsilon). \tag{7.27}$$

This is nothing but the Widom scaling hypothesis applied to the 1D Ising model. Here we can identify the power of b with the dimensionality d, and exponents y and x introduced earlier. However, the 1D Ising model is somewhat pathological as the x variable is not $T - T_c$ but e^{-4K}. This prevents us from going below T_c (since $T_c = 0$).

Problems

Problem 7.1 (From M. E. Fischer). *Consider a magnetic system near the critical point. The equation of state has the following form $h \approx aM(t + bM^2)^\theta$, where h is the magnetic field, M is the magnetization, $t = (T - T_c)/T_c$, $1 < \theta < 2$, and $a, b > 0$. (a) Find the critical exponents β, γ, and δ, where β is the order parameter (magnetization) exponent, γ is associated with the magnetic susceptibility, and δ determines the isothermal curve at the critical temperature. (b) Show that the exponents satisfy $\gamma = \beta(\delta - 1)$.*

(c) *Determine the scaling with magnetic field h of the Helmholtz free energy* $F(h,t)$ *at the critical temperature* $t = 0$.

Problem 7.2. *The finite-size scaling for an Ising ferromagnetic system takes the form*

$$f(b^y t, b^x h, b/L) = b^d f(t, h, L).$$

where f is the singular part of the free energy per site, $t = |T - T_c|/T_c$ *is the relative deviation away from the critical temperature, h is magnetic field, L is the linear size (length) of the system, and d is dimension of the system. Exactly at the critical point when* $t = 0$ *and* $h = 0$, *show that* (a) *Magnetization per spin* $m \propto L^{\Delta_1}$ *and determine the exponent* Δ_1 *in terms of x, y, and d.* (b) *Magnetic susceptibility* $\chi \propto L^{\Delta_2}$ *and also determine the exponent* Δ_2. (c) *Finally, find (or argue) the exponent* Δ_3 *for the quantity known as Binder's 4-th order cumulant*

$$\frac{\langle (\sum_i \sigma_i)^4 \rangle}{\langle (\sum_i \sigma_i)^2 \rangle^2} \propto L^{\Delta_3}.$$

Problem 7.3. *Consider the Landau theory for ferromagnetic phase transitions. The Gibbs free energy as a function of temperature T and total magnetization M is assumed to be*

$$G = (T - T_c)aM^2 + bM^4,$$

where a, b, and T_c *are some positive constants.* (a) *Give the corresponding expression for the Helmholtz free energy F as a function of temperature T and magnetic field h.* (b) *Let* $T = T_c$, *show that* $F(T_c, h) \propto h^{4/3}$. (c) *At* $h = 0$, $T < T_c$, *show that* $F(T, 0) \propto (T - T_c)^2$. (d) *Assuming a scaling form for the Helmholtz free energy*

$$F(t, h) = b^{-4} F(b^Y t, b^X h), \quad t = (T - T_c)/T_c,$$

and using the information obtained in (b) *and* (c), *determine the scaling exponents X and Y.*

Problem 7.4. (a) *The scaling of the singular part of the free energy per particle near a critical point is* $f(t, h) = b^{-d} f(b^y t, b^x h)$. *Similar equation holds for the correlation length as* $\xi(t, h) = b\xi(b^y t, b^x h)$. *Using the definition that* $\xi(t, 0) \propto t^{-\nu}$, *show that* $y = 1/\nu$. (b) *Show that the total susceptibility for an Ising model can be expressed as a d-dimensional integral of* *the correlation function:*

$$\chi = \frac{\partial M}{\partial h} \approx \frac{N}{k_B T} \int d^d\mathbf{r}\, G(r), \quad G(r) = \langle \sigma_i \sigma_{i+r} \rangle - \langle \sigma_i \rangle \langle \sigma_{i+r} \rangle.$$

(c) *The correlation function $G(r)$ satisfies the scaling relation*

$$G(r, \xi) = b^{-d+2-\eta} G(r/b, \xi/b).$$

Using this scaling relation and the result from (b), prove the Fisher scaling law, $\gamma = \nu(2-\eta)$, where the susceptibility exponent γ is defined by $\chi \propto t^{-\gamma}$.

Problem 7.5. *In the Kardar-Parisi-Zhang surface growth model, the height as a function of 1D position x and the time t follows the following scaling law:*

$$h(t, x) = b^\alpha h(b^{-z}t, b^{-1}x).$$

(a) *Show that the width of the growth interface in an interval of length L satisfies the scaling $W(L, t) \approx L^\alpha f(t/L^z)$, where the width is defined by*

$$W(L, t) = \left\langle \frac{1}{L} \int_0^L (h(x, t) - \bar{h}(t))^2 dx \right\rangle^{1/2},$$

where $\bar{h}(t)$ is the mean surface height at time t. (b) Determine the asymptotic behavior of $f(u)$ when $u \ll 1$ (short time, large system) and when $u \gg 1$ (long time, finite system).

Chapter 8

Monte Carlo Methods

The flavor of this chapter will be quite different from the earlier chapters. We will discuss numerical methods, in particular, the Monte Carlo method. The Monte Carlo method is one of the most widely used numerical methods in physics — from quantum field theory to condensed matter physics. It is an approach when there is no exact solution or when approximation methods fail. In this chapter, we first review the basics of the Monte Carlo method [Kalos and Whitlock (1986)]. We then go on to some applications of the Monte Carlo method in statistical physics problems [Landau and Binder (2015); Gubernatis et al. (2016)].

8.1 Random variables of specified distribution

The Monte Carlo method relies on the use of random numbers. Many computer programming languages have implemented a uniformly distributed random variable on the interval $[0, 1)$. We will denote such a variable ξ. In a computer program, each time a random number generator is called, a different value results. A common and simple method to generate such random values are the linear congruential method, by the recursion relation $x_{n+1} = (ax_n + c) \bmod m$, with a proper choice of the multiplier a, modulus m, and constant c, all of them integers. An example of a possible choice is $a = 6364136223846793005$, $m = 2^{64}$, $c = 1$ due to D. E. Knuth. A real value is obtained with $\xi = x_n/m$.

Assuming that such a uniform real number between 0 and 1 is available, we discuss methods of obtaining other distributions. Let us say that we want to generate a random integer i taking value 1 with probability p_1, value 2 with probability $p_2, \dots$, and value n with probability p_n. The sum of the probabilities is one,

$$\sum_{i=1}^{n} p_i = 1. \tag{8.1}$$

The random variable i distributed according to probability p_1, p_2, ..., p_n can be generated by taking one random number ξ and deciding a value for i: if $\xi \leq p_1$, then set $i = 1$; if $p_1 < \xi \leq p_1 + p_2$, then $i = 2, \ldots$, if $\sum_{k=1}^{m-1} p_k < \xi \leq \sum_{k=1}^{m} p_k$, then $i = m$, etc. The most common case is $n = 2$. A special case $p_1 = p_2 = \cdots = p_n = 1/n$ (the random variable i takes value 1 to n with equal probability) can be obtained with a simple operation $i = \lfloor n\xi \rfloor + 1$, where $\lfloor \ \rfloor$ denotes the floor or truncation to an integer.

Let's look at another example. We want to sample the geometric distribution $p_i = 2^{-i}$, $i = 1, 2, \ldots$. According to the above method, we should set $i = 1$ if $\xi < 1/2$; $i = 2$ if $1/2 < \xi < 1/2 + 1/4$; and in general $i = k$ if $\sum_{j=1}^{k-1} 2^{-j} < \xi < \sum_{j=1}^{k} 2^{-j}$. We can express the condition more concisely as $i = -\lfloor \ln \xi / \ln 2 \rfloor + 1$.

8.1.1 *Continuous distribution*

Consider the distribution of a single random variable x taking real values in some domain. Its probability density is $p(x)$. Probability density is defined such that the probability that the random variable x takes values between x and $x + dx$ is $p(x)\,dx$. The cumulative distribution function $F(x)$ is defined by the probability that the random variable x is less than or equal to a given value x_0,

$$P(x \leq x_0) = F(x_0) = \int_{-\infty}^{x_0} p(x)\,dx. \tag{8.2}$$

Since $F(x)$ is a probability, we have $0 \leq F(x) \leq 1$. The distribution function is a nondecreasing function of its argument.

The random variable x with probability density $p(x)$ can be generated with the formula $x = F^{-1}(\xi)$, where $F^{-1}(\xi)$ is the inverse function of the distribution function $F(x)$, and ξ is a uniformly distributed random number.

Let's prove that x is indeed distributed according to $p(x)$. Since there is a one-to-one correspondence between x and ξ, the probability for x taking values between x and $x + dx$ is the same as for ξ between ξ and $\xi + d\xi$.

Since ξ is uniformly distributed, the probability in this interval is just $d\xi$. And since $\xi = F(x)$, we have the probability that x is between x and $x + dx$ as $1 \cdot d\xi = dF(x) = F'(x) \, dx = p(x) \, dx$.

Example 1. Generate x according to the exponential distribution $p(x) = e^{-x}$, with $x \geq 0$. The distribution function $F(x) = \int_{-\infty}^{x} p(x) \, dx = \int_{0}^{x} e^{-x} \, dx = 1 - e^{-x}$, $x \geq 0$. The inverse function is $x = -\ln(1 - \xi)$. One can also equivalently generate the exponential distribution with the formula $x = -\ln \xi$ (Why is this so?).

Example 2. Generate the Gaussian distribution $p(x) = \frac{1}{\sqrt{2\pi}} e^{-\frac{1}{2} x^2}$, $-\infty < x < \infty$. It is helpful to generate a pair of Gaussian random numbers, since the inverse of the Gaussian distribution function (error function) cannot be expressed by elementary functions. The joint distribution of x and y is $p(x, y) = \frac{1}{2\pi} e^{-\frac{1}{2}(x^2 + y^2)}$, $-\infty < x, y < \infty$. Let us introduce polar coordinates $r = \sqrt{x^2 + y^2}$ and $\theta = \tan^{-1}(y/x)$, we can rewrite the probability density as $p(x, y) \, dx \, dy = \frac{1}{2\pi} e^{-\frac{1}{2} r^2} r \, dr \, d\theta$. From the above expression, we see that θ is distributed uniformly between 0 and 2π; and r is distributed according to $re^{-r^2/2}$. Thus the distribution function for r is $F_r(r) = \int_0^r re^{-\frac{1}{2} r^2} dr = 1 - e^{-\frac{1}{2} r^2} = \xi_1$. Solving for r, $r = \sqrt{-2\ln(1 - \xi_1)}$. We can replace $1 - \xi_1$ by ξ_1 since it does not change the probability distribution. The random variable θ can be generated by $\theta = 2\pi\xi_2$. And finally, x and y can be generated by the formula

$$x = r\cos\theta = \sqrt{-2\ln\xi_1} \, \cos(2\pi\xi_2), \tag{8.3}$$

$$y = r\sin\theta = \sqrt{-2\ln\xi_1} \, \sin(2\pi\xi_2). \tag{8.4}$$

Here ξ_1 and ξ_2 are two independent random numbers uniformly distributed between 0 and 1. This procedure is known as the Box-Muller method. One can easily rescale the standard Gaussian distribution to get a Gaussian distribution with mean a and variance σ^2 by $\sigma x + a$.

The rejection method is a method which does not rely on the analytic expression of inverse distribution functions. But it is less efficient. Assuming that $a \leq x \leq b$, and $0 \leq p(x) \leq c$, the number $x = a + \xi_1(b - a)$ is accepted if $c\xi_2 < p(x)$; otherwise a new value x is drawn and the condition is tested again.

The method of using the inverse distribution function is useful only if the inverse can be found easily. And it works only if one can reduce the problem to one random variable. In later sections, we'll study methods (importance sampling) of generating arbitrary distributions involving many random variables. But we will first review numerical integration by the Monte Carlo method.

8.2   Monte Carlo evaluation of finite-dimensional integrals

Let $X = (x_1, x_2, \ldots, x_d)$ be a point in a d-dimensional space, and $dX = dx_1 dx_2 \cdots dx_d$ the volume element. We want to evaluate the integral over the domain Ω,

$$G = \int_\Omega g(X) \, p(X) \, dX, \qquad (8.5)$$

where $p(X)$ is interpreted as the probability density satisfying $p(X) \geq 0$, $\int_\Omega p(X) \, dX = 1$. The basic Monte Carlo method of calculating the integral is to draw a set (sequence) of random values $X_1, X_2, \ldots, X_N$ distributed according to the probability $p(X)$, and to form the arithmetic mean

$$G_N = \frac{1}{N} \sum_{i=1}^{N} g(X_i). \qquad (8.6)$$

The quantity G_N is approximately G, and we say G_N is an estimator of G. It is crucial to understand why $p(X_i)$ is not multiplied to $g(X_i)$. If X_i were equally spaced in the domain Ω, we would include $p(X_i)$. Since X_i is distributed according to $p(X_i)$ — there are more points where $p(X_i)$ is large — we have already taken it into account. Very often the probability $p(X)$ itself is a uniform distribution $p(X) = \text{const}$ in the domain Ω. Other times, we need other forms of the distribution, e.g., in equilibrium statistical mechanics, we often like to sample the Boltzmann distribution, $p(X) \propto e^{-E(X)/k_B T}$. Splitting the integrand $g(X)p(X)$ into a probability $p(X)$ and a quantity $g(X)$ is somewhat arbitrary and is for the convenience of computation. However, the choice of $p(X)$ does influence the statistical accuracy of Monte Carlo estimates.

8.2.1   *Fundamental theorems*

We'll state two important theorems which form the theoretical basis of the Monte Carlo method.

(1) **Law of Large Numbers**: the arithmetic mean G_N converges with probability one to the expectation value G. In mathematical notation $P\{\lim_{N \to \infty} G_N = G\} = 1$. This theorem guarantees the convergence

of Monte Carlo estimates to the correct answer in the limit of infinitely many points. But it does not tell us how fast it converges. The next theorem tells us more.

(2) **Central Limit Theorem**: Let's define the variance as $\sigma^2 = \int_\Omega (g(X) - G)^2 p(X) \, dX = \langle g^2 \rangle - \langle g \rangle^2$, and a normalized value (also a random variable) $t_N = \sqrt{N} \, (G_N - G)/\sigma$, then $\lim_{N \to \infty} P\{a \le t_N < b\} = \int_a^b \frac{1}{\sqrt{2\pi}} e^{-\frac{1}{2}t^2} \, dt$. In other words, G_N is a random variable distributed according to a Gaussian distribution with mean G and variance σ^2/N (with uncorrelated samples X_i) for sufficiently large N. As $N \to \infty$, the observed G_N turns up in an ever narrower interval near G and one can predict the probability of deviations in units of σ. The observed G_N is within one *standard error* (i.e., $\sigma/\sqrt{N}$) of G 68.3% of the time, within two standard errors 95.4% of the time, and within three standard errors 99.7% of the time. This is due to the property of the Gaussian distribution.

8.2.2 *Error of Monte Carlo calculations*

From the central limit theorem, we have $G = G_N +$ error. The error of Monte Carlo estimates for G is $\epsilon \approx \frac{\sigma}{\sqrt{N}}$.

From the definition of σ, we can also estimate the error ϵ as well as the value G itself. However, it is better to use the following *unbiased* estimator for σ^2.

$$\sigma^2 \approx \frac{N}{N-1} \left\{ \frac{1}{N} \sum_{i=1}^{N} g^2(X_i) - \left(\frac{1}{N} \sum_{i=1}^{N} g(X_i) \right)^2 \right\}. \tag{8.7}$$

Monte Carlo error decreases as $1/\sqrt{N}$, as the number of samples (points) N increases. Since N is proportional to computer CPU time T, error $\epsilon \propto T^{-1/2}$. Let's compare the efficiency of Monte Carlo method with that of deterministic method of quadrature (e.g. trapezoidal or Simpson rule) [Press *et al.* (1992)]. The deterministic methods typically have error $\epsilon \propto h^k$ due to discretization, where h is the grid size. The computer time goes as $T \propto h^{-d}$ in d dimensions when the grid size is decreased. Thus error goes as $\epsilon \propto \frac{1}{T^{k/d}}$. For Simpson rule of numerical integration, $k = 4$. We see that for a one-dimensional integral, the error decreases much faster than that of the Monte Carlo method. However, as the dimension of the integral increases, the Simpson rule becomes worse than the Monte Carlo method for a fixed amount of computer time.

In conclusion, standard numerical quadrature is very good at low dimensions. The Monte Carlo method is superior for higher dimensional integrations.

8.3   Importance sampling (Metropolis algorithm)

In the Monte Carlo computation of integration outlined in the above section, we assumed that we could generate X with probability $P(X)$. We have studied methods of generating one-dimensional distributions by inverting the distribution function and by rejection. However, transformation methods only work if one can invert the function analytically, and the rejection method is inefficient. In higher dimensions, we need other methods of generating the distribution. Importance sampling is a method of generating probability distributions efficiently on a computer.

8.3.1   *Markov chain*

A Markov chain [Norris (1997)] is a mathematical notion for a random walk in the space of X. Remember that $X = (x_1, x_2, \ldots, x_d)$ is a point in a d-dimensional space. We'll sometimes refer X as a state of the system. For mathematical convenience, we will consider only discrete space points (think of the Ising model). We start from some initial point X_0; the next point X_1 is generated according to certain transition probabilities. In this way a sequence of random points X_0, X_1, $\ldots$, X_N is generated. These points are required to appear with the probability $P(X)$. The rule to generate next point X_{i+1} given the point X_i is described by the transition probability $P(X_{i+1}|X_i) = W(X_i \to X_{i+1})$, where $W(X_i \to X_{i+1})$ is the transition probability from state X_i to X_{i+1}, and $P(X_{i+1}|X_i)$ is the probability that the system is in state X_{i+1} given that the system was in state X_i, namely, the conditional probability. We have $W(X_i \to X_{i+1}) \geq 0$, $\sum_{X_{i+1}} W(X_i \to X_{i+1}) = 1$. Since W is a probability, it satisfies the usual constraint of probability — it must be positive and sum to one. The method of importance sampling is to choose a proper transition probability W in such a way that the distribution $P(X)$ is the desired distribution we want to generate. More precisely, let $P_0(X)$ be some probability distribution, which is the initial distribution of the states. The new distribution after one step is then $P_1(X) = \sum_{X'} P_0(X')W(X' \to X)$. In general we have

$$P_k(X) = \sum_{X'} P_{k-1}(X')W(X' \to X). \tag{8.8}$$

The theorems of stochastic processes tell us that the limit for large k exists and is unique (with certain conditions) $\lim_{k \to \infty} P_k(X) \to P(X)$. And $P(X)$ satisfies the equation $P(X) = \sum_{X'} P(X')W(X' \to X)$. We say that $P(X)$ is invariant under the transition of W, or that $P(X)$ is the equilibrium distribution for the given transition probabilities.

To have a better, intuitive understanding of the above equations, one may think of the probability as population of certain particles. Imagine that the particles can be in a number of (quantum-mechanical) states denoted by X. Initially at time 0, the particles are distributed according to $P_0(X)$. Each particle has a tendency to go to other states. For example, if the particle is at state X', the probability that it will go to state X is $W(X' \to X)$. Thus, we have (number of particles times the probability) $P_0(X')W(X' \to X)$ particles going to state X. After a unit time, the population of particles at state X, $P_1(X)$, is the sum of the number of particles going from any states X' to the state X. These transitions of particles will eventually make the population steady in any states — we have reached the equilibrium.

In a computer simulation, typically only one state X is stored in memory. The transition changes the state. The distribution of the state is also given by $P_k(X)$. By this we mean that if we do the same simulation many times starting with X_0 distributed according to $P_0(X)$, the outcome after k steps will be distributed according to $P_k(X)$.

The sequence of points X_i generated by the random walk is correlated, since the next point is generated based on the knowledge of the previous points. This is in contrast to the simple sampling method, e.g., the inverse distribution function method, where the sequence of points is uncorrelated. Each point is generated afresh. The penalty due to the correlation is that the statistical error will be larger than the formula $\sigma/\sqrt{N}$ implies, which is valid only if the points are uncorrelated.

8.3.2 *Ergodicity*

We mentioned the existence and uniqueness of the equilibrium distribution. But we skipped over the condition for it. The condition is the following: starting from any state X, it is possible to get to any another state X' after a finite number of transitions. We call such a Markov chain ergodic (or regular, irreducible). We can view the transition probability W as a matrix. Let $W^n(X, X')$ be the matrix element of the n-th power of W. The ergodic condition means that the matrix element at every entry (X, X') is nonzero for all $n > n_0$.

8.3.3   Detailed balance

How do we choose W to get the prescribed P? It is sufficient that W satisfies

$$P(X')W(X' \to X) = P(X)W(X \to X'), \qquad (8.9)$$

where P is the known distribution. If W satisfies this equation and it is ergodic, then the Markov chain will generate X according to the probability distribution $P(X)$. It is possible that W does not satisfy the detailed balance condition but still generates the distribution P. That's why we say that the above condition is a sufficient condition.

Let's prove that the distribution is invariant under the transition W satisfying the detailed balance condition. Summing over the variable X' and using the normalization condition for W, we have

$$\sum_{X'} P(X')W(X' \to X) = \sum_{X'} P(X)W(X \to X') =$$
$$P(X) \sum_{X'} W(X \to X') = P(X). \qquad (8.10)$$

8.3.4   Metropolis algorithm

The importance Markov chain Monte Carlo sampling for a specific distribution $P(X)$ goes as follows: starting from some initial state (configuration) X_0, a sequence of states will be generated by a Markov chain through the transition probabilities $W(X_i \to X_{i+1})$. The starting state may be a unique state or may be generated at random with a certain distribution P_0. The new state X_{i+1} is based on the old state X_i. The new state is generated in such a way so that the probability that the system is in state X_{i+1} is just $W(X_i \to X_{i+1})$, knowing that the system was in state X_i. Hopefully we can generate this probability distribution easily. Otherwise we might have generated $P(X)$ directly in the first place. It is sufficient that W satisfies the detailed balance condition and the Markov chain is ergodic. Metropolis algorithm refers to a specific choice of W:

$$W(X \to X') = T(X \to X') \min\left(1, \frac{P(X')}{P(X)}\right), \quad X \neq X', \qquad (8.11)$$

with $T(X \to X') = T(X' \to X)$. Matrix T can be any distribution but must be symmetric. We almost always use $T(X \to X') = $ const if X' is inside a certain region around X, and 0 otherwise. The probability that the state does not change, $W(X \to X)$, is determined by the normalization condition $\sum_{X'} W(X \to X') = 1$.

Let's describe more concretely the algorithm.

Metropolis Algorithm: Generate a sequence $X_0, X_1, X_2, \ldots, X_i, \ldots$, with a probability distribution $P(X_i)$.

0) Specify an initial state X_0, set $i = 0$.
1) Generate a new state $\overline{X}_{i+1}$ according to $T(X_i \to \overline{X}_{i+1})$.
2) Accept the new state with probability $\min\big(1, P(\overline{X}_{i+1})/P(X_i)\big)$. That is, $X_{i+1} = \overline{X}_{i+1}$ if a random number between 0 and 1 is less than $P(\overline{X}_{i+1})/P(X_i)$; retain the old state as the $(i+1)$-th state, $X_{i+1} = X_i$, otherwise.
3) Set $i = i + 1$, go to 1).

Expectation values (or averages) can be estimated by $\langle A(X) \rangle = \sum_X A(X)P(X) \approx \frac{1}{N} \sum_{i=M+1}^{M+N} A(X_i)$. It is important that we throw away the first M points in the average. We must wait for the system to approach equilibrium before we use the points. The reason for this is that the system is distributed according to $P_k(X)$, and we want the distribution $P(X)$. $P_k(X)$ will converge to $P(X)$ for sufficiently large k. This convergence is usually fast. For efficiency consideration, usually we do not use every X_i even if we have reached equilibrium, since they are highly correlated. We calculate $A(X_i)$ only every certain number of iterations or steps.

8.3.5 Example of the Metropolis algorithm — Ising model for magnets

Let us take the Ising system as an example. On an L^d hypercubic lattice, we put a spin σ_i at each lattice site i. For each configuration denoted by $\{\sigma\} = (\sigma_1, \sigma_2, \sigma_3, \ldots)$ (the collection of all the spins, with $\sigma_i = \pm 1$), the system has a definite energy given by $H(\{\sigma\}) = -J \sum_{\langle i,j \rangle} \sigma_i \sigma_j - h \sum_i \sigma_i$, where J is the coupling constant, h is proportional to the external magnetic field. The first summation runs over the nearest neighbors only, i.e., each bond is summed once. The second summation is over all the lattice sites. We assume J is positive. The model describes a ferromagnet.

According to statistical-mechanical description, the system will not be in a definite state. Due to thermal fluctuation, the system will have some probability distribution among the states. If the temperature T is a fixed parameter, we have the famous Boltzmann distribution (canonical distribution) $P(\{\sigma\}) \propto \exp(-\frac{H(\{\sigma\})}{k_B T})$, where $k_B \approx 1.38 \times 10^{-23}$ joule/kelvin is the Boltzmann constant. Our goal is to calculate average quantities such as energy, magnetization, correlation functions, etc. The internal energy per spin is defined by $u = \frac{1}{L^d} \langle H \rangle = \frac{1}{L^d} \sum_{\{\sigma\}} H(\{\sigma\}) P(\{\sigma\})$, and the magnetization is defined as $m = \frac{1}{L^d} \langle M \rangle = \frac{1}{L^d} \sum_{\{\sigma\}} |\sum_i \sigma_i| P(\{\sigma\})$. The summation is over all possible states. Since each site can have two states (spin up and spin down), we have 2^{L^d} states for a system of linear size L in d dimensions.

A few other quantities are also useful in Ising model simulations. The specific heat $c = \partial u / \partial T$ and susceptibility $\chi = \partial m / \partial h$ can be calculated from the fluctuations $c = \frac{1}{k_B T^2 L^d}[\langle H^2 \rangle - \langle H \rangle^2]$, $k_B T L^d \chi = \langle M^2 \rangle - \langle M \rangle^2$. Here the angular brackets $\langle \cdots \rangle$ denote average over the Boltzmann distribution. And $M = |\sum_i \sigma_i|$ is the absolute value of the total magnetization of a particular state. The fourth-order cumulant $g = \frac{1}{2}\left[3 - \langle M^4 \rangle / \langle M^2 \rangle^2\right]$, is useful in determining the critical temperature.

A Monte Carlo simulation of the Ising model involves the following steps:

0) Initialize σ_i at each site with arbitrary spin values, e.g., all spin up, or up or down with equal probability. The complete set of spins $\{\sigma\}$ is the abstract state X discussed in the previous sections.

1) Choose a site i at random and propose to flip the spin at that site, $\sigma_i' = -\sigma_i$. The proposed state $\overline{X}$ is the one with one spin at location i flipped.

2) Calculate the energy increment $\Delta E = H(\{\sigma'\}) - H(\{\sigma\}) = 2J\sigma_i \sum_{\substack{\text{neighbors} \\ \text{of } i}} \sigma_j + 2h\sigma_i$.

3) Accept the proposed state as the new state if a uniformly distributed random number (between 0 and 1) is less than $\exp(-\Delta E / k_B T)$; retain the old state as the new state otherwise.

4) Go to 1).

One Monte Carlo sweep will be defined as performing the above basic single-spin flip L^d number of times. After one Monte Carlo sweep, each site on average has tried once to flip. One can also go through the lattice in sequence, and try to flip each spin at the site. But choosing sites at random seems better.

8.4   Analysis of the Ising model Monte Carlo data

We have introduced what is known about the Ising model and then presented the general picture of second-order phase transitions in earlier chapters. We now discuss how to analyze the data obtained from Monte Carlo simulation.

8.4.1 Finite-size scaling

The Ising model in two and higher dimensions has a phase transition from ferromagnetic phase to paramagnetic phase. Many quantities near the critical temperature T_c obey a power law. The magnetization behaves as $m \propto (T_c - T)^\beta$, $T < T_c$, $T \to T_c$, and the specific heat diverges $c \propto |T - T_c|^{-\alpha}$, $T \to T_c$. The susceptibility also diverges $\chi \propto |T - T_c|^{-\gamma}$, $T \to T_c$. For the two-dimensional Ising model, the critical exponents are $\beta = 1/8$, $\alpha = 0$, and $\gamma = 7/4$. The specific heat exponent $\alpha = 0$ is consistent with $c \propto \ln |T - T_c|$.

The theory and analytical results outlined above apply to infinitely large systems. In the language of statistical mechanics, they are results in the thermodynamic limit. However, a computer simulation is always performed on finite lattices. This is acceptable in most of the range of parameters of the model. The effects of finite sizes show up when the physics gets interesting — near the critical point the finite-size effects become important.

In fact, the finite-size effect is not necessarily a drawback. It can be utilized to determine the ratios of critical exponents. According to finite-size scaling theory, the values of thermodynamic quantities at the critical point vary with sizes as follows

$$c(T_c) \propto L^{\alpha/\nu}, \quad m(T_c) \propto L^{-\beta/\nu}, \quad \chi(T_c) \propto L^{\gamma/\nu}, \tag{8.12}$$

where the exponent ν is called the correlation length exponent ($\nu = 1$ for the two-dimensional Ising model). The specific heat and susceptibility have peaks near the critical point. The location of the peak depends on the system size. Finite-size scaling theory predicts that the peak will shift according to $T_{peak}(L) - T_c \propto L^{-1/\nu}$. This is one way to determine the location of the critical temperature T_c as well as the exponent ν. However, because we have three fitting parameters (T_c, ν, and the proportionality constant) it is not easy to get an accurate value for ν.

The finite-size scaling theory gives a general description of the temperature and size dependencies of the thermodynamic quantities. In general, if T is the only variable in the thermodynamic limit, then a quantity, say, the susceptibility, has the following finite-size dependence $\chi(T, L) = L^{\gamma/\nu} \tilde{\chi}((T - T_c)L^{1/\nu})$. The function $\tilde{\chi}(x)$ is called the (universal) scaling function. The finite-size dependence of the fourth-order cumulant is particularly interesting: $g(T, L) = \tilde{g}((T - T_c)L^{1/\nu})$. It becomes size independent at T_c. Below T_c it increases with sizes and above T_c it decreases with sizes. We can use this property to determine T_c. The use of finite-size scaling together with efficient cluster algorithm discussed later and modern data analysis techniques have yielded exceedingly accurate critical exponents for the three-dimensional Ising model [Ferrenberg *et al.* (2018)].

8.5   Critical slowing down

If one compares the simulation of percolation and that of the Ising model, the simulation of percolation is computationally more efficient in that each configuration generated is independent of the previous configuration. In a site percolation problem [Stauffer and Aharony (1994)], each site on a lattice is occupied with probability p or empty with probability $1 - p$. In the Ising model simulation, or more generally any simulation using the Metropolis algorithm, the next configuration depends on the previous one. Due to this correlation between Monte Carlo steps, the formula for the statistical error $\epsilon \approx \sigma/\sqrt{N}$ underestimates the true error. The error formula should be replaced by $\epsilon \approx \sigma\sqrt{\frac{1+2\tau}{N}}$. That is, the error is larger by a factor $\sqrt{1 + 2\tau}$ than the uncorrelated data. The quantity τ is called the correlation time, which is roughly the number of Monte Carlo steps needed to generate independent configurations. We can compute the correlation time τ as follows: define the time-dependent correlation function,

$$f(t) = \frac{\langle A(t')A(t'+t)\rangle - \langle A\rangle^2}{\langle A^2\rangle - \langle A\rangle^2}, \tag{8.13}$$

where time t is measured in terms of Monte Carlo sweeps — time is passed by one unit when one updates the system by one Monte Carlo sweep. And $A(t')$ is the quantity at steps t' whose error we want to estimate. The integrated correlation time is defined by $\tau = \sum_{i=1}^{\infty} f(t)$. In an actual numerical estimate of τ we cannot sum up to infinity due to statistical errors in $f(t)$. Typically we truncate the sum or extrapolate the correlation function to infinity.

Now we are ready to explain the concept of critical slowing down. Just like many other quantities near the critical point, the correlation time also becomes singular at the critical point. On an infinite lattice, it obeys a power-law

$$\tau \propto |T - T_c|^{-\nu z}. \tag{8.14}$$

Here ν is the correlation length exponent and z is the dynamic critical exponent. This phenomenon that the correlation time becomes very large is called critical slowing down. It not only happens in computer simulations but also appears in real experiments. On a finite system of linear size L, the correlation time will, of course, not go to infinity. But it will grow with size as $\tau \propto L^z$, at $T = T_c$. Substituting this result into the error formula, we find that $\epsilon \approx \sigma L^{z/2}/N^{1/2}$ for large system near T_c. For the two-dimensional Ising model with the Metropolis algorithm, $z \approx 2.1$, we see that typically increasing the size leads to a larger error in the quantity to be calculated.

8.6 Cluster Monte Carlo algorithms

The Metropolis algorithm makes changes locally one site at a time. This is the cause of the critical slowing down. There are algorithms that change the states by a group of clusters. The advantage of the cluster algorithms is that they are faster, not necessarily in computer time per Monte Carlo step, but in terms of dynamics. That is, it has a much smaller value of z.

8.6.1 *Swendsen-Wang algorithm*

The Swendsen-Wang multi-cluster algorithm is closely related to percolation. One starts with a spin configuration $\{\sigma\}$ and generates a percolation configuration based on the spin configuration by the rule described below, see also Fig. 8.1. Then the old spin configuration is forgotten and a new spin configuration $\{\sigma'\}$ is generated based on the percolation configuration. The rule is such that the detailed balance is satisfied. Or more generally, the transition leaves the equilibrium probability invariant. The following is the algorithm for the nearest neighbor Ising model without magnetic field ($h = 0$). It can be generalized to other models with or without magnetic field.

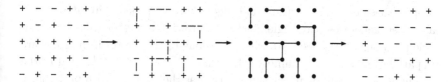

Fig. 8.1 The Swendsen-Wang algorithm: starting from an Ising spin configuration shown on the left, bonds are laid down with probability $p = 1 - e^{-2J/(k_B T)}$ between the nearest neighbor pairs of same spins. Then, after erasing the spins, a percolation bond configuration results. Each of the clusters, defined by the sites connected by bonds (including these of isolated sites) is given a random value of up or down with equal probability. One step is finished.

Swendsen-Wang Algorithm:

0) Start with some arbitrary state $\{\sigma\}$.

1) Go through each nearest neighbor connection of the lattice, create a bond between the two neighboring sites i and j with probability $p = 1 - e^{-2J/(k_B T)}$, only if the spins are the same, $\sigma_i = \sigma_j$. Never put a bond between the sites if the spin values are different. We are creating a bond percolation configuration with probability p on a subset of lattice sites where the spins are either all pointing up or down.

2) Identify clusters as a set of sites connected by bonds, or isolated sites. Two sites are said to be in the same cluster if there is a connected path of bonds joining them. Every site has to belong to one of the clusters. After the clusters are found, each cluster is assigned a new Ising spin chosen with equal probability between $+1$ and -1. The old spin values now can be 'forgotten'. And all the sites in a cluster now take the value of the spin assigned to the cluster.

3) One Monte Carlo step is finished. Repeat 1) for the next step.

The performance of the algorithm in terms of correlation time in comparison with the Metropolis algorithm is remarkable, see Fig. 8.2. Recall that the dynamic critical exponent $z \approx 2$ for the Metropolis algorithm almost independent of the dimensionality. The Swendsen-Wang algorithm in one dimension gives $z = 0$ (τ approaches a constant as $T \to 0$). For the two-dimensional Ising model, the dynamic critical exponent z is less than 0.3 or possibly zero (but $\tau \propto \ln L$). In three dimensions it is about 0.5. At and above four dimensions it is 1.

8.6.2 *Wolff single-cluster algorithm*

In the Swendsen-Wang algorithm, we generated many clusters and then flipped these clusters. In the Wolff algorithm one picks a site at random, and then generates one single cluster by growing a cluster from the seed. The neighbors of the seed site will also belong to the cluster if the spins are in parallel and a random number is less than $p = 1 - e^{-2J/(k_B T)}$. That is, the neighboring site will be in the same cluster as the seed site with probability p if the spins have the same sign. If the spins are different, neighboring site will never belong to the cluster. Neighbors of each new site in the cluster are tested for membership. This testing of membership is performed on pair of sites (forming a nearest neighbor bond) not more than once. The recursive process will eventually terminate. The spins in the cluster are

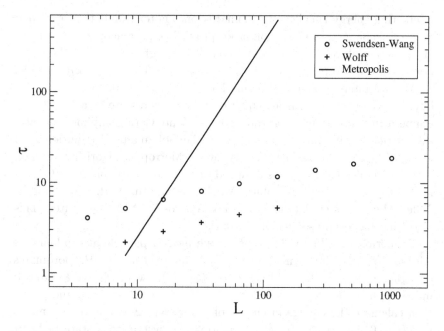

Fig. 8.2 The correlation time τ vs. system linear size L for the two-dimensional Ising model at the critical temperature T_c. The straight line indicates a slope of $z \approx 2.167$ of the dynamic critical exponent for the 2D Ising model with single-spin flip dynamics.

turned over with probability 1. This single-cluster algorithm appears more efficient than the multi-cluster one.

8.7   Other applications of the Metropolis algorithm

The Metropolis algorithm and similar algorithms have wide applications in physics. Examples are quantum system and quantum field theory, fluids, macro-molecules and polymer systems. In this section, we'll describe briefly how to simulate fluid.

8.7.1   *Simple fluid*

Let's imagine that we have a (two-dimensional) box of dimension $L \times L$. Inside, we put a certain number of particles. Depending on the density of the particles, we might get a fluid state or a solid state. We assume that the particles can be described as classical point particles. The particles interact with each other via potential energy $v(r) = 4\,\epsilon[(\frac{\sigma}{r})^{12} - (\frac{\sigma}{r})^{6}]$. This form of interacting energy is called the Lennard-Jones potential. It is a very good model for noble gases like argon or krypton. The parameter ϵ is the depth of the potential well, and σ can be thought as roughly the size of the atom. For argon the two parameters have the value $\sigma \approx 3.5\,\text{Å}$ and $\epsilon/k_B \approx 118\,\text{K}$.

We are interested in the calculation of thermodynamic quantities like internal energy, specific heat, equation of state (pressure as a function of temperature and volume), entropy, etc. A Monte Carlo sampling generates the particle configurations according to the Boltzmann distribution $P \propto \exp(-E/k_B T)$. This can be achieved by the Metropolis algorithm. In fact, the Metropolis algorithm was designed just for such a problem in 1953, by N. Metropolis, A. W. Rosenbluth, M. N. Rosenbluth, A. H. Teller, and E. Teller [Metropolis *et al.* (1953)] on a Los Alamos MANIAC computer. This algorithm has been used widely since then.

To perform a Monte Carlo move, one picks a particle at random and tries to move it within a small region, say a square of side a with coordinates changing as $x \to x + \left(\xi_1 - \frac{1}{2}\right)a$, $y \to y + \left(\xi_2 - \frac{1}{2}\right)a$, where ξ_1 and ξ_2 are two independent uniformly distributed random numbers between 0 and 1. We then calculate the change in energy of the system ΔE, due to this move. If $\Delta E \leq 0$, i.e., if the move would bring the system to a state of lower energy, we allow the move and put the particle in the new location. If $\Delta E > 0$, we allow the move with probability $\exp(-\Delta E/k_B T)$; i.e., we take a random number ξ between 0 and 1, and if $\xi < \exp(-\Delta E/k_B T)$, we move the particle to the new position. Otherwise we return it to its old position. Note, however, whether we moved the particle or not, we always count it as one Monte Carlo attempt. Once we have an equilibrium distribution, it is possible to calculate the equation of state by the virial theorem (relating pressure with certain averages of force, see problem set 8.7).

Problems

Problem 8.1. *In the Buffon needle experiments, one throws needles of length L onto strips of spacings d. Prove the probability of the needle intersecting the equally spaced lines is $p = 2L/(\pi d)$, where L is the length of the needle, and d is strip spacing, and $L < d$.*

Problem 8.2. *Suppose that ξ_1 and ξ_2 are two random variables uniformly distributed between 0 and 1. What is the probability distribution of the new variable which is a sum of the two: $x = \xi_1 + \xi_2$? (Hint: determine the* *cumulative distribution function* $F(x)$ *of the new variable, then* $p(x) = dF(x)/dx$.

Problem 8.3. *Design an algorithm to generate a random variable* x *with the probability density* $p(x) = x\exp(-x)$, $x \geq 0$, *using a transformation from uniformly distributed random numbers* ξ_i *(Hint: verify that* $-\ln(\xi_1\xi_2)$ *will do the job).*

Problem 8.4. *(a) Give the transition matrix* W *(an* 8×8 *matrix) of a Monte Carlo simulation for a three-spin Ising system using the Metropolis flip rates. A spin is chosen among the three with equal probability. The Hamiltonian (energy) of the model is*

$$H(\sigma) = -J(\sigma_1\sigma_2 + \sigma_2\sigma_3 + \sigma_3\sigma_1) - h(\sigma_1 + \sigma_2 + \sigma_3), \quad \sigma_i = \pm 1,$$

where J *and* h *are some positive constants, and assuming* $J = h$ *for simplicity. (b) What is the left eigenvector of* W *with eigenvalue 1, i.e., find* p, *such that* $p = pW$. *Discuss the meaning of* p.

Problem 8.5. *A sequence of uniformly distributed random numbers* ξ_1, $\xi_2, \ldots$, *from the interval* $[0,1)$ *are given. Determine a transform using the random numbers through the help of Cholesky decomposition,* $LL^T = A^{-1}$, *such that the resulting random vector* x *is given by the Gaussian distribution* $\propto \exp(-\frac{1}{2}x^T A x)$. *Here* A *is a symmetry, positive definite matrix and superscript* T *stands for matrix transpose.*

Problem 8.6. *Consider Monte Carlo sampling of a one-dimensional Ising model with energy function* $H(\sigma) = -J\sum_{i=1}^{N}\sigma_i\sigma_{i+1}$ *in the canonical ensemble. We assume periodic boundary condition, that is,* $\sigma_{N+i} = \sigma_i$. N *is some arbitrary natural number larger than 1, and the coupling parameter* $J > 0$, $\sigma_i = \pm 1$. *The Gibbs sampling is slightly different from the Metropolis algorithm, as follows: first, pick a site* $1 \leq i \leq N$ *at random. Compute* $p = x/(x + x^{-1})$, *here* $x = e^{\beta(\sigma_{i-1} + \sigma_{i+1})}$, $\beta = 1/(k_B T)$. *Then set* σ_i *as* $+1$ *with probability* p, *or* -1 *with probability* $1-p$. *(a) Write down the associated transition matrix* $W(\sigma \to \sigma')$. *(b) Show that the detailed balance condition is satisfied, that is,* $P(\sigma)W(\sigma \to \sigma') = P(\sigma')W(\sigma' \to \sigma)$. *Here* $P(\sigma) = e^{-\beta H(\sigma)}/Z$, *and* Z *is the partition function. (c) Give a complete Python code to realize this algorithm.*

Problem 8.7. *Using the canonical ensemble with the configurational partition function, show that the equation of state for a fluid system with* central-force pair potentials can be expressed as

$$pV = Nk_BT + \frac{1}{d}\left\langle \sum_{i<j}(\mathbf{r}_i - \mathbf{r}_j) \cdot \mathbf{F}_{ij} \right\rangle,$$

here d is the dimension of the system, $\mathbf{r}_i$ is the position vector of particle i, and $\mathbf{F}_{ij}$ is the force acting on particle i from particle j.

Chapter 9

Brownian Motion — Langevin and Fokker-Planck Equations

9.1 Introduction

Having studied equilibrium statistical mechanics where the probability distributions of the systems are assumed or postulated, in this chapter, we discuss particularly successful methods [Kubo *et al.* (1992); Zwanzig (2001); Pottier (2010)] where the question of a system approaching equilibrium can be addressed. We shall first discuss the phenomenon of Brownian motion, and how such phenomenon can be described from an atomic point of view. We focus on a single particle and write down a Newtonian equation for it with random noise, which is the Langevin equation. In a Fokker-Planck approach, the noise is eliminated in favor of the probability distribution. The partial differential equation for the probability distribution is the Fokker-Planck equation. With a single particle, we can then show that in the long time or in steady state, the distribution follows Maxwell's distribution, just as we have assumed in an equilibrium canonical ensemble.

9.2 Brownian motion

In 1827 Robert Brown observed that pollens floating in water make irregular motions. Such irregular motions appear in many contexts, such as a mirror mounted on suspension fiber, or random current in electric circuits. Its origin was not well-understood until Einstein gave a satisfactory molecular theory in 1905.

If we have a macroscopically large number of "Brownian" particles but is still sufficiently dilute, the movement of the particles satisfies Fick's law:

$$j = -D\frac{\partial n}{\partial x}. \tag{9.1}$$

Here $j = nv$ is the particle current density, i.e., the number of particles passing through a surface per unit area per unit time. v is the average velocity of the particles. n is the number of particles per unit volume (concentration). For simplicity we assume the particles flow in the x direction. Fick's law asserts that the current density is proportional to the concentration gradient. The proportionality constant D is called the diffusion constant. By a simple dimensional analysis, the units of D must be, in SI units, m^2/s. D is typically a small number in SI units, for example, the diffusion constant of gas in air is of the order $10^{-6} m^2/s$; NaCl in water about $1.2 \times 10^{-10} m^2/s$, and solid in another solid $10^{-12} m^2/s$. Since the total number of particles is a conserved quantity, the concentration and current density is related by the continuity equation,

$$\frac{\partial n}{\partial t} + \nabla \cdot j = 0. \tag{9.2}$$

Together with Fick's law, we obtain the diffusion equation:

$$\frac{\partial n}{\partial t} = D\frac{\partial^2 n}{\partial x^2}. \tag{9.3}$$

An insightful suggestion of Einstein is that this equation should also be valid if we have only one particle. In that case, since $\int n\, dx = 1$, n is interpreted as the probability density. Let us put one particle at the origin at time $t = 0$, then by symmetry $\langle x \rangle = \int x\, n\, dx = 0$. So on average the particle is not moving anywhere. We are then interested in the mean-square displacement,

$$\langle x^2 \rangle = \int_{-\infty}^{+\infty} x^2 n(x, t)dx. \tag{9.4}$$

The time change of the mean-square displacement can be computed with the help of the diffusion equation [Zwanzig (2001)],

$$\frac{\partial \langle x^2 \rangle}{\partial t} = \int x^2 D\frac{\partial^2 n}{\partial x^2}dx$$

$$= D\left[x^2\frac{\partial n}{\partial x}\bigg|_{-\infty}^{+\infty} - \int 2x\frac{\partial n}{\partial x}dx \right]$$

$$= 2D. \tag{9.5}$$

Here we have performed integration by parts two times, and assumed that n and its gradient are 0 at $\pm\infty$. The above equation implies $\langle x^2 \rangle = 2Dt +$ const. For sufficiently long time, we have $\langle x^2 \rangle \approx 2Dt$. This relation is known as the Einstein relation.

The diffusion equation can be solved using well-known methods such as the Fourier transform. If the initial density at time $t = 0$ is given by $n_0(x)$, then density profile is given then at a later time as

$$n(x,t) = \int p(x,t|x')n_0(x')\,dx', \quad p(x,t|x') = \frac{1}{\sqrt{4\pi Dt}} e^{-\frac{(x-x')^2}{4Dt}}. \quad (9.6)$$

We can find the kernel of the integral representation by solving a special initial condition problem of a delta function distribution, $n(x,0) = N\delta(x)$, here N is the total number of particles. Let us represent the solution by a Fourier representation in space, as

$$n(x,t) = \int_{-\infty}^{+\infty} \frac{dk}{2\pi} \tilde{n}(k,t)e^{ikx}. \quad (9.7)$$

Substituting this expression into the diffusion equation, we turn the partial differential equation into an ordinary differential equation with the wave-vector k as a parameter,

$$\frac{\partial \tilde{n}}{\partial t} = -Dk^2\tilde{n}. \quad (9.8)$$

The solution is a simple exponential relaxation in time, $\tilde{n} = Ne^{-Dk^2t}$. At $t = 0$, $\tilde{n}$ is a constant. This gives the delta function as the initial concentration distribution. The result in real space is obtained by performing a Gaussian integral after completing the square:

$$n(x,t) = N\int_{-\infty}^{+\infty} \frac{dk}{2\pi} e^{-Dk^2t+ikx} = N\frac{1}{\sqrt{4\pi Dt}} e^{-\frac{x^2}{4Dt}}. \quad (9.9)$$

The integral kernel is obtained when we set $N = 1$ and shift the argument to $x - x'$. A notable feature of this solution is that we can only predict the future, not the past. As if $t < 0$, the square root in the prefactor of the Gaussian function becomes meaningless.

9.2.1 Random walk model

We can give a simple model to demonstrate the fact that the mean-square displacement is proportional to time t in a Brownian motion. The Brownian particle can be modeled as performing a random walk at each discrete time

$i\Delta t$, moving to the left or right with equal probability. Then at time t the position of the particle is at

$$x(t) = \sum_{i=1,2,\ldots,t/\Delta t} \xi_i a, \qquad (9.10)$$

where a is the step length, $\xi_i = \pm 1$. We assume the choices of left or right are completely random and uncorrelated, thus,

$$\langle \xi_i \rangle = 0, \quad \langle \xi_i \xi_j \rangle = \delta_{ij}. \qquad (9.11)$$

From this, we find that on average, the particle is not moving anywhere, $\langle x(t) \rangle = 0$. And the variance of the displacement is

$$\langle x(t)^2 \rangle = \sum_{i=1}^{N} \sum_{j=1}^{N} \langle \xi_i \xi_j \rangle a^2 = Na^2 = 2Dt. \qquad (9.12)$$

Here $N = t/\Delta t$, and we have defined the diffusion constant as $D = a^2/(2\Delta t)$.

If a particle is moving with a constant velocity, $x \propto t$, we will call this 'ballistic' motion, while $x \propto \sqrt{t}$ is diffusive motion. The diffusive motion has much to do with thermal fluctuation and equilibrium. This connection will be made quantitative in the next section where we discuss the Langevin equation.

9.3   Langevin equation

In fluid dynamics, if we have a particle of radius a moving with velocity v, the particle experiences a drag force which slows the particle down. If the velocity is not too high, the drag force is proportional to the velocity, given by Stokes' law, $F = -6\pi a \eta v$, where η is the viscosity of the fluid. The viscosity is defined as the force per unit area needed in parallel plate geometry such that the two plates move with unit relative velocity per unit separation distance, i.e., $F/A = \eta \, dv/dz$.

For a small particle in a fluid, in addition to the drag force, there is also a random force which tries to speed up the motion of the particle. The random force $R(t)$ is due to the existence of molecules (say water molecules) which form the fluid. These two effects lead to the equation

$$m\frac{dv}{dt} = -m\gamma v + R(t), \qquad (9.13)$$

where m is the mass of the particle, and $m\gamma = 6\pi a \eta$. This is known as the Langevin equation. The random force $R(t)$ is a difficult concept. Its precise specification needs the concept of probability. Here we demand that its averages satisfy

$$\langle R(t) \rangle = 0, \quad \langle R(t)R(t') \rangle = C\delta(t - t'). \tag{9.14}$$

The first equation says that, since it is random, on average, the force is 0. However, the correlation of the force is not zero, but is given by a delta function. We call such a random force white noise. The constant C is not arbitrary, but must be fixed by the requirement that after a long time, the system reaches equilibrium and equal-partition theorem is valid.

An equation of type (9.13) is called a stochastic differential equation, as it contains a partially specified function $R(t)$. To appreciate the meaning of the random force $R(t)$, we consider how this equation can be solved on a computer numerically. Given an initial time t, a small-time increment Δt later the velocity is obtained by integrating from t to $t + \Delta t$ on both sides of the equation,

$$v(t + \Delta t) - v(t) \approx -\gamma v(t)\Delta t + \frac{1}{m} \int_t^{t+\Delta t} R(t')dt'. \tag{9.15}$$

Here on the right-side we can replace the velocity integral by its approximate value, but such replacement for the random force is not allowed as it is not a smooth function. The property of the random force is now given by an integral,

$$\xi = \int_t^{t+\Delta t} R(t')dt'. \tag{9.16}$$

The average value of ξ is 0 since $R(t')$ is, the variance of ξ can be computed from the correlation function of $R(t)$, given $\langle \xi^2 \rangle = C\Delta t$. This random noise is realized on a computer by generating Gaussian random numbers with mean 0 and variance $C\Delta t$. Thus, the random variable ξ fluctuates with a magnitude of order $\sqrt{C\Delta t}$ proportional to the square root of Δt instead of linear to it.

The moments of $R(t)$ are given by (gaussian or Ornstein-Uhlenbeck process) properties of their averages in Eq. (9.14), not its specific value of R at a given time t. As a result, a physically meaningful result can only be obtained through average properties. Let us first consider the average velocity. Since the average of the random force is 0, the average velocity satisfies a simple equation

$$m\frac{d\langle v \rangle}{dt} = -m\gamma\langle v \rangle. \tag{9.17}$$

This gives an exponential relaxation toward zero, $\langle v \rangle = v_0 e^{-\gamma t}$, where v_0 is the initial velocity at time $t = 0$. This solution tells us even though we can give an initial velocity in a certain direction, it will lose the direction after a time of order $1/\gamma$. The motion of the Brownian particle will eventually be random.

Next, we consider the correlations between velocities at different times, i.e., the average $\langle v(t)v(t') \rangle$. To this end, we need to solve the differential equation formally, with $R(t)$ as given. We can then take the average of the required correlation function. A standard way of solving such an inhomogeneous differential equation is the method of 'variation of a constant'. First, dividing the mass m throughout, and we set $R(t) = 0$ and the equation becomes $dv/dt = -\gamma v$, the solution is the same as before for the average, $v(t) = A e^{-\gamma t}$. We now set the constant A as a function of t, and we assume a solution of the form,

$$v(t) = A(t)e^{-\gamma t}. \tag{9.18}$$

This form is substituted back into the original equation, after a cancellation of the $-\gamma v$ term, we find $dA(t)/dt = R(t)e^{\gamma t}/m$. Upon integration, we find

$$A(t) = A(0) + \int_0^t \frac{R(\tau)}{m} e^{\gamma \tau} d\tau. \tag{9.19}$$

The velocity is then given by

$$v(t) = v_0 e^{-\gamma t} + \int_0^t \frac{R(\tau)}{m} e^{-\gamma(t-\tau)} d\tau. \tag{9.20}$$

In computing the correlation function $\langle v(t)v(t') \rangle$, we obtain four terms, the constant term independent of $R(t)$ is $v_0^2 e^{-\gamma(t+t')}$, the cross terms that are first order in R average to 0, the $R(\tau)R(\tau')$ term can be replaced by the delta function, we obtain:

$$\langle v(t)v(t') \rangle = v_0^2 e^{-\gamma(t+t')} + \int_0^t d\tau \int_0^{t'} d\tau' \frac{e^{-\gamma(t-\tau+t'-\tau')}}{m^2} C\delta(\tau - \tau')$$

$$= v_0^2 e^{-\gamma(t+t')} + \frac{C}{2\gamma m^2}[e^{-\gamma|t-t'|} - e^{-\gamma(t+t')}]. \tag{9.21}$$

In performing the two-dimensional integral over the rectangle $[0, t] \times [0, t']$ with a delta function on the diagonal, it is important to pay attention to the two cases of $t < t'$ and $t > t'$. The final result valid for any $t, t' > 0$

can be combined as the absolute value $|t - t'|$. After a long time such that both t and t' are large but the difference is finite, we can ignore the first and last terms since they are exponentially small. As a result, time translational invariance is restored, and we have $\langle v(t)v(t')\rangle \approx \frac{C}{2\gamma m^2}e^{-\gamma|t-t'|}$. In particular, at equal times $t = t'$, and after a long time, we demand that the equal-partition theorem holds, $\frac{1}{2}m\langle v^2\rangle = \frac{1}{2}k_BT$, this fixes the constant C as

$$C = 2m\gamma k_BT. \tag{9.22}$$

This is known as the classical version of the fluctuation-dissipation theorem. The random fluctuation of noise must be related to equilibrium properties of the system, in order for the system to reach equilibrium after a long time.

Another property of the velocity of the solution of the Langevin equation we can show is that it is a gaussian random variable. We compute the 'characteristic function' $\langle e^{i\xi v(t)}\rangle$. This requires the correlation functions of the random noise $R(t)$ with any number of products of $R(t)$. To this end, we make assumption to the averages of products of $R(t)$ as follows. (1) Any odd number of $R(t)$ averages to 0. (2) Any even number of R follows "Wick's theorem":

$$\langle R(t_1)R(t_2)\cdots R(t_n)\rangle = \sum_{\text{all possible pairs}} \langle R(t_i)R(t_j)\rangle \cdots \langle R(t_kR(t_l)\rangle. \tag{9.23}$$

For example, if we have 4 terms, $\langle 1, 2, 3, 4\rangle$, the three possible pairings are $\langle 1, 2\rangle\langle 3, 4\rangle$, $\langle 1, 3\rangle\langle 2, 4\rangle$, and $\langle 1, 4\rangle\langle 2, 3\rangle$. Using this property, the moment generating function or the characteristic function can be computed and the distribution of v is determined by only two quantities, average and variance. As a result, we find [Kubo *et al.* (1992)]

$$\langle e^{i\xi v(t)}\rangle = \exp\left[i\xi v_0e^{-\gamma t} - \frac{k_BT}{2m}(1 - e^{-2\gamma t})\xi^2\right]. \tag{9.24}$$

This has the general form of a generating function of a gaussian random variable with mean $\langle v(t)\rangle = v_0e^{-\gamma t}$ and variance $\langle (v(t) - \langle v(t)\rangle)^2\rangle = (1 - e^{-2\gamma t})k_BT/m$. The logarithm of the above expression is called the cumulant generating function, and the coefficients of power of $i\xi$ are called the cumulants. All cumulants higher than second order are zero for a gaussian process.

9.4   Einstein relation

9.4.1   Langevin's argument

In one of the annus mirabilis papers of Einstein of 1905, he treated Brownian motion and derived a famous formula named after him:

$$D = \frac{k_B T}{6\pi a \eta},$$ (9.25)

here D is the diffusion constant of the Brownian particle defined according to $\langle x^2 \rangle = 2Dt$ (for large t), a is the radius of the Brownian particle, and η is the viscosity of the fluid that the particle is immersed in. Three years later, Langevin (1908) gave a simpler derivation of the same result starting from his equation, Eq. (9.13). A key insight of Langevin is that the position of the particle and random force $R(t)$ are uncorrelated, $\langle x(t)R(t) \rangle = 0$. Writing the Langevin equation in terms of position x instead of velocity v,

$$\frac{d^2 x}{dt^2} = -\gamma \frac{dx}{dt} + \frac{R(t)}{m},$$ (9.26)

multiplying the equation by x and then taking the average, the noise term drops out and we obtain

$$\left\langle x \frac{d^2 x}{dt^2} \right\rangle = -\gamma \langle x \dot{x} \rangle.$$ (9.27)

Here we have introduced the short-hand notation $dx/dt = \dot{x}$. We note the identities $dx^2/dt = 2x\dot{x}$, and $d^2 x^2/dt^2 = 2\dot{x}^2 + 2x \, d^2 x/dt^2$. Thus, the equation can be put in a form

$$\left\langle \frac{1}{2} \frac{d^2 x^2}{dt^2} - \dot{x}^2 \right\rangle = -\frac{\gamma}{2} \frac{d}{dt} \langle x^2 \rangle.$$ (9.28)

Evoking the equipartition theorem, $\langle \dot{x}^2 \rangle = k_B T/m$, and defining $d\langle x^2 \rangle/dt = z$, we find a first order differential equation in z,

$$\frac{dz}{dt} + \gamma z = \frac{2k_B T}{m},$$ (9.29)

which can be solved to give $d\langle x^2 \rangle/dt = z = 2k_B T/(\gamma m) + A e^{-\gamma t}$, here A is the integration constant. Integrating one more time, and taking the large t limit, the exponential term drops, we obtain $\langle x^2 \rangle = 2Dt = 2t k_B T/(\gamma m)$. Recalling the relation between the friction coefficient and Stokes' law, $\gamma = 6\pi a \eta/m$, we recover the Einstein relation.

9.4.2 *Velocity correlation and diffusion constant*

We can make a connection between the mean-square displacement and the velocity correlation, thus also the diffusion constant. Assuming the particle is at the origin at time 0, we have

$$x(t) = \int_0^t v(\tau)d\tau. \tag{9.30}$$

Taking the square and then averaging over random noise, we obtain

$$\langle x(t)^2 \rangle = \int_0^t d\tau \int_0^t d\tau' \langle v(\tau)v(\tau') \rangle. \tag{9.31}$$

We assume that a steady state is reached (i.e., after a long time) so that we assume that the correlation function is time translationally invariant, $\langle v(\tau)v(\tau') \rangle = \langle v(0)v(\tau' - \tau) \rangle$. We make a change of variable by $s = \tau' - \tau$. We find

$$\langle x(t)^2 \rangle = \int_0^t d\tau \int_{-\tau}^{t-\tau} ds \langle v(0)v(s) \rangle = 2 \int_0^t ds(t - s)\langle v(0)v(s) \rangle. \tag{9.32}$$

In obtaining the last expression, we have switched the order of integration. As a result, the τ-integration can be performed explicitly. We also used the fact that the correlation is an even function of the variable s. Finally, when the observation time t is much larger than the velocity correlation time, $t - s \approx t$, and we have

$$D = \int_0^\infty ds \langle v(0)v(s) \rangle = \frac{k_B T}{m\gamma}. \tag{9.33}$$

Here we have used the explicit velocity correlation function from the previous section. This again recovers the Einstein relation. In our derivation, we have used the long-time limit. If t is small, we could have found $\langle x(t)^2 \rangle \propto t^2$, which is a ballistic behavior.

9.4.3 *Mobility and diffusion*

The Brownian particle on average will not move anywhere. As we have seen earlier, its average velocity will decay to 0 if not further disturbed. In order for the Brownian particle to move with a constant average velocity, we must drive it with a constant force f. Thus, we can revise the Langevin equation under external force as

$$m\frac{dv}{dt} = -m\gamma v + f + R(t). \tag{9.34}$$

This equation describes, for example, a particle under gravity, the problem of sedimentation, or a charged particle moving in an electric field. In steady state, the average velocity is then proportional to the applied force,

$$\langle v \rangle = \frac{1}{m\gamma} f = \mu f. \tag{9.35}$$

Here the proportionality constant μ is defined as the mobility. Comparing it with the diffusion constant in the above subsection, we find

$$\mu = \frac{D}{k_B T}. \tag{9.36}$$

This equation is also called "Einstein's" relation. It is an instance of the fluctuation-dissipation theorem. The diffusion constant is related to the velocity fluctuations in equilibrium, while the mobility is related to energy dissipation since it is related the drag coefficient γ. The mobility represents a response of the system when it is driven externally out of equilibrium.

9.5   Fourier method, Wiener-Khinchin theorem

If we are interested in the quantities after a long time, stationary states are established so that the correlation functions are functions of time difference only. In such a situation, it is useful and economical to solve the Langevin equation in the frequency domain. Then a powerful theorem known as Wiener-Khinchin theorem tell us how to relate the power spectrum (to be defined below) and the correlation in real time. First, we introduce the notation of the Fourier transform and its inverse,

$$a(\omega) = \int_{-\infty}^{+\infty} z(t) e^{i\omega t} dt, \tag{9.37}$$

$$z(t) = \int_{-\infty}^{+\infty} a(\omega) e^{-i\omega t} \frac{d\omega}{2\pi}. \tag{9.38}$$

We consider $z(t)$ as representing some physical quantity such as velocity, and assume it is real, then the Fourier transform has the symmetry $a^*(\omega) = a(-\omega)$. The correlation function is defined by

$$\phi(t) = \langle z(t + t_0) z(t_0) \rangle, \tag{9.39}$$

which we have assumed to be dependent only on the time difference of the two z's. We will define the Fourier transform of $\phi(t)$ as

$I(\omega) = \int_{-\infty}^{+\infty} \phi(t)e^{i\omega t}dt$. Then consider

$$\langle a(\omega)a^*(\omega')\rangle = \int dt \int dt' \langle z(t)z(t')\rangle e^{i(\omega t - \omega' t')}$$

$$= \int dt'' \phi(t'')e^{i\omega' t''} \int_{-\infty}^{+\infty} dt e^{i(\omega - \omega')t}$$

$$= I(\omega)2\pi\delta(\omega - \omega'). \tag{9.40}$$

In the above, we have made a change of variable from t' to $t'' = t - t'$, and the fact that the integral of last factor produces 2π times a Dirac δ function. The frequency dependent function $I(\omega)$ will be called the power spectrum of a. The above result is the content of the Wiener-Khinchin theorem.

Let us apply this theorem to the solution of the Langevin equation again in steady state. First, the noise correlation is $\langle R(t)R(t')\rangle = C\delta(t-t')$, thus its Fourier transform is the constant C, we will denote it by $I_R(\omega) = C$. The theorem implies that $\langle \tilde{R}(\omega)\tilde{R}(\omega')^*\rangle = 2\pi C\delta(\omega - \omega')$. Here $\tilde{R}(\omega)$ is the Fourier transform of the time domain $R(t)$. The Langevin equation itself can be Fourier transformed to give us

$$-i\omega\tilde{v}(\omega) = -\gamma\tilde{v}(\omega) + \frac{\tilde{R}(\omega)}{m}. \tag{9.41}$$

There the tilde versions are Fourier transforms of the original variables, e.g., $\tilde{v}(\omega) = \int_{-\infty}^{+\infty} v(t)e^{i\omega t}dt$. As a result, the differential equation is reduced to an algebraic equation. The solution is readily found to be $\tilde{v} = \frac{1}{m}\tilde{R}/(-i\omega + \gamma)$. Applying the Wiener-Khinchin theorem and use the analytic solution for $\tilde{v}$, we have

$$\langle \tilde{v}(\omega)\tilde{v}^*(\omega')\rangle = \frac{\langle \tilde{R}(\omega)\tilde{R}^*(\omega')\rangle}{m^2(-i\omega + \gamma)(i\omega' + \gamma)} = I_v(\omega)2\pi\delta(\omega - \omega'). \tag{9.42}$$

Comparing this to the similar formula for the noise, we find

$$I_v(\omega) = \frac{C}{m^2|-i\omega + \gamma|^2}. \tag{9.43}$$

The correlation function in real time can be found by an inverse transform.

$$\langle v(t)v(0)\rangle = \int_{-\infty}^{+\infty} \frac{Ce^{-i\omega t}}{m^2(\omega^2 + \gamma^2)}\frac{d\omega}{2\pi} = \frac{C}{2m^2\gamma}e^{-\gamma|t|}. \tag{9.44}$$

To obtain the final expression, we need to apply the residue theorem for complex functions, $\oint f(z)dz = 2\pi i \sum_j \text{Res} f(z_j)$. The residue inside the closed contour at z_j is the coefficient a_{-1} of the (simple) pole $a_{-1}/(z - z_j)$.

There are two simple poles located at $\pm i\gamma$. In order for the exponential term to decay to 0 at infinity, we close a contour with a semi-circle in the lower half plane if $t > 0$, or upper half plane if $t < 0$. At time $t = 0$, the expression gives us the mean-square value of the velocity, $\langle v^2 \rangle$. Applying the equipartition theorem, $\frac{1}{2} m \langle v^2 \rangle = k_B T / 2$, we again obtain $C = 2m\gamma k_B T$.

9.6   Fokker-Planck equation

An alternative point of view of the stochastic process is to consider the probability distribution. For the Brownian particle, we consider the probability distribution $P(v, t)$ of the velocity v at time t. More precisely, $P(v, t)dv$ gives the probability of finding the Brownian particle at velocities in the interval v to $v + dv$ at time t. There are at least two methods to derive the equation. The first is that of Kramers-Moyal expansion [Reif (1965); Risken (1989)]. Here we consider how this probability changes due to the motion of the particle. This is related to the moments of the velocity change in the short time limit,

$$M_n = \lim_{\tau \to 0} \frac{1}{\tau} \langle (v(\tau) - v)^n \rangle, \quad v = v(\tau = 0). \tag{9.45}$$

For a gaussian process, only the first two moments are nonzero, thus the expansion terminates to second order. The Fokker-Planck equation is

$$\frac{\partial P}{\partial t} = -\frac{\partial}{\partial v}(M_1 P) + \frac{1}{2}\frac{\partial^2}{\partial v^2}(M_2 P). \tag{9.46}$$

Here, we give a derivation following Zwanzig [Zwanzig (2001)]. Our starting point is again the Langevin equation, $dv/dt = -\gamma v + R(t)/m$. We first work out an equation for $P(v, t)$ in a formal way for a given realization of the random noise $R(t)$. This equation then contains the random noise. The next step is to take an ensemble average of the equation over the noise, and show it results in the Fokker-Planck equation.

Since $P(v, t)$ is the probability density and the total probability must be 1, so we have $\int_{-\infty}^{+\infty} P(v, t)dv = 1$. A differential form of this conservation of probability is

$$\frac{\partial P}{\partial t} + \frac{\partial}{\partial v}(\dot{v} P) = 0. \tag{9.47}$$

This equation is the continuity equation much like the charge conservation in electricity and magnetism, or the conservation of probability in phase space, discussed in Chap. 2. $\dot{v}P$ gives the current of the probability, and $\dot{v} = -\gamma v + R(t)/m$ is the right-hand side of the Langevin equation. This equation contains the random noise, so it is not the Fokker-Planck equation. We transform this equation into a form that is suitable to do the ensemble average over the noise. Specifically, we prefer to have an equation that involves two R's so that we can use the fact that the correlation of R is a delta function. To this end, we convert the equation into an integral form. Formally we write the equation as

$$\frac{\partial P}{\partial t} = \hat{L}P + f, \tag{9.48}$$

here $\hat{L}P = \partial(\gamma v P)/\partial v$ is a differential operator in v space acting on P, and $f = -\partial(R(t)P/m)/\partial v$ still depends on P. Equations of this form can be solved again with the variation of a constant method. First, by assuming that $f = 0$, we find $P(v,t) = e^{t\hat{L}}P(v,0)$. However, since $f \neq 0$, we replace $P(v,0)$ by some arbitrary function $A(v,t)$, and find that it satisfies $e^{t\hat{L}}\dot{A} = f$. The inverse of the exponential operator is $e^{-t\hat{L}}$, thus we can solve for A and obtain the solution for P as

$$P(v,t) = e^{t\hat{L}}P(v,0) - \int_0^t d\tau\, e^{(t-\tau)\hat{L}} \frac{\partial}{\partial v}\left(\frac{R(\tau)}{m}P(v,\tau)\right). \tag{9.49}$$

We note that $P(v,t)$ appears on both the left and right side, so it is still an equation for $P(v,t)$. If we take the equation and put it back to the right-hand side, we see we can expand $P(v,t)$ in terms of the initial distribution $P(v,0)$ and power of the noise $R(t)$. We have obtained a formal solution for P. However, we prefer still an equation instead of a formal solution, we take Eq.(9.49) and plug it back into f in Eq. (9.48). By doing this, we obtain two R's multiplied (at different time arguments). We are now ready to take the ensemble average over the noise. First, the single R term multiplied by the distribution drops out, $\langle R(t)e^{t\hat{L}}P(v,0)\rangle = 0$. We can rearrange the position of $R(t)$ as it commutes with the differential operator $\hat{L}$, so we obtain

$$\langle f \rangle = \frac{\partial}{\partial v}\left\langle \frac{R(t)}{m}\int_0^t d\tau e^{(t-\tau)\hat{L}} \frac{\partial}{\partial v}\frac{R(\tau)}{m}P(v,\tau)\right\rangle$$

$$= \frac{1}{m^2}\frac{\partial}{\partial v}\int_0^t d\tau e^{(t-\tau)\hat{L}} \frac{\partial}{\partial v}\langle R(t)R(\tau)P(v,\tau)\rangle$$

$$= \frac{C}{2m^2}\frac{\partial^2}{\partial v^2}\langle P(v,t)\rangle. \tag{9.50}$$

In arriving at the final result, we have assumed the validity of Wick's theorem, so that $\langle R(t)R(\tau)P(v,\tau)\rangle = \langle R(t)P(\tau)\rangle\langle P(v,\tau)\rangle$, and the integration ends right at the middle of a δ-function, thus the factor of $1/2$ instead of 0 or 1. Using the result that $C = 2m\gamma k_B T$, and for notational simplicity, drop the angular brackets on P, we find the Fokker-Planck equation as

$$\frac{\partial P(v,t)}{\partial t} = \gamma \frac{\partial v P(v,t)}{\partial v} + \frac{\gamma k_B T}{m}\frac{\partial^2 P(v,t)}{\partial v^2}. \tag{9.51}$$

In deriving the above equation, we see that two conditions are needed, (1) R is gaussian in the sense of Wick's theorem, and (2) R is δ-correlated (white noise).

9.6.1   *Solution of the Fokker-Planck equation*

With the Fokker-Planck equation at our disposal, we can ask how does the Brownian particle equilibrate after a long time. First, let us consider the steady state where equilibrium has already been reached. Then the distribution should be independent of time t, so $\partial P/\partial t = 0$. Thus, $P(v)$ must satisfy

$$\frac{\partial}{\partial v}(vP) + \frac{k_B T}{m}\frac{\partial^2 P}{\partial v^2} = 0, \tag{9.52}$$

or $vP + (k_B T/m)\partial P/\partial v = $ const. The constant must be equal to 0, for if that is not the case, then if we integrate over v from $-\infty$ to $+\infty$, the left side is a finite quantity, but right-hand side would be infinite. So we can simplify to, $dP/P = -mvdv/(k_B T)$, or integrating one more time, we find $P(v) \propto e^{-\frac{1}{2}mv^2/(k_B T)}$, which is nothing but the Maxwell distribution for the velocity.

To find the approach to the Maxwell distribution, we need to solve the partial differential equation, which in general is difficult. Fortunately for the current equation, we can find the integration factor and transform the Fokker-Planck equation into a diffusion equation [Reif (1965)], for which we do have a solution for a special initial condition. The transformation is found by first omitting the second derivative term, and considering

$$\frac{\partial P}{\partial t} - \gamma v\frac{\partial P}{\partial v} = \gamma P. \tag{9.53}$$

Multiplying by λ which is assumed to be a function of v and t, and demanding that it is proportional to a total differential, i.e, $\frac{\partial P}{\partial t}dt + \frac{\partial P}{\partial v}dv = dP$, we identify

$$dt = \lambda, \quad dv = -\lambda\gamma v, \quad \text{and} \quad dP = \lambda\gamma P. \tag{9.54}$$

We find $dv/dt = -\gamma v$ and $dP/dt = \gamma P$; the solutions are $v = ue^{-\gamma t}$ and $P = Qe^{\gamma t}$. We thus change the constants to variables and propose the following transformation $v \to u$ and $P \to Q$:

$$P(v,t) = Q(u,t)e^{\gamma t}, \quad u = ve^{\gamma t}. \tag{9.55}$$

We need to express the partial derivatives of the old variables in terms of the new variables. Using chain rule, we find

$$\frac{\partial P}{\partial t} = \frac{\partial Q}{\partial t}e^{\gamma t} + \gamma ue^{\gamma t}\frac{\partial Q}{\partial u} + Q\gamma e^{\gamma t}, \tag{9.56}$$

$$\frac{\partial P}{\partial v} = e^{\gamma t}\frac{\partial Q}{\partial u}\frac{\partial u}{\partial v} = e^{2\gamma t}\frac{\partial Q}{\partial u}, \tag{9.57}$$

$$\frac{\partial^2 P}{\partial v^2} = e^{2\gamma t}\frac{\partial^2 Q}{\partial u^2}\frac{\partial u}{\partial v} = e^{3\gamma t}\frac{\partial^2 Q}{\partial u^2}. \tag{9.58}$$

Finally, one more transformation is needed, $d\theta = e^{2\gamma t}dt$, or $\theta = (e^{2\gamma t} - 1)/(2\gamma)$. Plugging all the transformations into the original Fokker-Planck equation, we obtain the diffusion equation

$$\frac{\partial Q}{\partial \theta} = \tilde{D}\frac{\partial^2 Q}{\partial u^2}, \tag{9.59}$$

here we define $\tilde{D} = \gamma^2 D = \gamma k_B T/m$. A special solution of the diffusion equation is the gaussian function,

$$Q(u,\theta) = \frac{1}{\sqrt{4\pi\tilde{D}\theta}}e^{-\frac{(u-u_0)^2}{4\tilde{D}\theta}}, \tag{9.60}$$

which corresponds to a delta-function initial condition, $Q \to \delta(u - u_0)$ as $\theta \to 0$. Transforming back to the original function and variables, with some simplification of the expression we find

$$P(v,t|v_0) = \left[\frac{m}{2\pi k_B T(1 - e^{-2\gamma t})}\right]^{1/2} \exp\left[-\frac{m(v - v_0e^{-\gamma t})^2}{2k_B T(1 - e^{-2\gamma t})}\right]. \tag{9.61}$$

It is a Gaussian distribution for the velocity with a mean $\langle v \rangle = v_0e^{-\gamma t}$ and a variance $\langle v^2 \rangle - \langle v \rangle^2 = (1 - e^{-2\gamma t})k_B T/m$, consistent with the result from the Langevin equation consideration. After a long time, $t \to \infty$, the distribution approaches the Maxwell distribution.

Although the example we have given so far is the simple 1D Langevin equation, the idea can be generalized in many ways for different systems. The central concept here is that we can focus on parts of the system of interest and treat other degrees of freedom through a phenomenological random noise. In so doing we have reduced the complexity of our problem and through the noise, the system of interest reaches thermal equilibrium after a long time. Such a viewpoint can also be used to study thermodynamics of small systems. The approach is valid and can be rigorously derived (e.g., the Caldeira-Leggett model), provided that the 'environment' varies at a much faster time scale than the system. In the standard Brownian particle problem, the Brownian particle is much heavier than the fluid molecules, so that the assumption that the environment is memoryless is well-justified. The Langevin and the associated Fokker-Planck equation can be generalized to many variables. Consider the dynamic variables given by a column vector $X = (x_1, x_2, \ldots, x_N)^T$, with an equation of motion of the form

$$\dot{X} = F(X) + g(X)\xi, \quad \langle \xi(t)\xi^T(t') \rangle = 2D\delta(t - t'), \quad (9.62)$$

Here $F(X)$ is a vector and $g(X)$ is an $N \times N$ matrix, and ξ is a column vector with white noise. The diffusion constant now is a matrix. If $g(X)$ is a function of the vector X, we call it a multiplicative noise; care must be taken to interpret the meaning of the stochastic differential equation (Ito vs. Stratonovich calculus). Using similar steps as for the one degree of freedom problem, one can show that the Fokker-Planck equation for the probability distribution $P(X, t)$ is [Lindenberg and West (1990)]

$$\frac{\partial P}{\partial t} = -\frac{\partial^T}{\partial X}\left(F(X)P\right) + \left(\frac{\partial}{\partial X}\right)^T gD\left[g^T \frac{\overset{\leftrightarrow}{\partial}}{\partial X} P\right]. \quad (9.63)$$

Here we should view $\frac{\partial}{\partial X}$ as a column vector in the component i of x_i space, and the last derivative acts both on P as well as on g^T.

9.7 Supplementary reading — A microscopic model of Brownian motion

In Sec. 9.3, we put down the Langevin equation by some physical considerations and good intuition. Indeed, it is possible to give a more 'rigorous' derivation of the Langevin equation. In so doing, we also understand the limitations of the Langevin equation. We consider one big particle of mass M moving in a harmonic potential and immersed in a lot of smaller particles. For simplicity, we assume the smaller particles are harmonic oscillators and interact with the big particle by harmonic springs. The Hamiltonian of the full system takes the form:

$$H = \frac{P^2}{2M} + \frac{1}{2}KX^2 + \sum_j \left(\frac{p_j^2}{2m_j} + \frac{1}{2}m_j\omega_j^2 \Big(q_j - \frac{u_j X}{m_j\omega_j^2} \Big)^2 \right). \tag{9.64}$$

This is known as the Caldeira-Leggett model (see, e.g., [Pottier (2010)], Supplement 10B). Here the first two terms are the kinetic and potential energies of the big particle. P and X are the momentum and position, K is the force constant of the big particle. The last terms summed over index j are the smaller particles' kinetic and potential energies. The masses of the smaller particles are m_j with intrinsic oscillation frequency ω_j for the j-th particle. u_j is the coupling parameter to the j-th mode. There are interacting terms proportional to $q_j X$ when the square is expanded. The specific form of the last term is to guarantee translational invariance in the final result.

The idea of a derivation of the Langevin equation goes as follows. We first solve the equations of motion for the environment or bath (the small particles), and put the formal result back into the equation of motion of the big particle. As a result, we see that due to the interacting terms, we find a convolution-in-time term related to the environment and an additional term we will call noise. The property of the noise is fixed by assuming that initially the environment is in thermal equilibrium. Finally, we assume that the number of oscillators for the environment goes to infinity, and they vary at a much faster time scale. By a so-called Markov approximation, we obtain the Langevin equation.

The coupled equations of motion, recalling Hamilton's equations of motion, and eliminating the momentum to get a Newtonian form, are:

$$M\ddot{X} = -(K + \Delta K)X + \sum_j u_j q_j, \quad \Delta K = \sum_j \frac{u_j^2}{m_j\omega_j^2}, \tag{9.65}$$

$$m_j\ddot{q}_j = -m_j\omega_j^2 q_j + u_j X. \tag{9.66}$$

The solution of the second equation can be expressed as

$$q_j(t) = q_j^0(t) - u_j \int_{t_0}^t g_j(t - t')X(t')dt', \tag{9.67}$$

where $q_j^0(t) = q_j(t_0)\cos(\omega_j(t - t_0)) + \frac{p_j(t_0)}{m_j\omega_j}\sin(\omega_j(t - t_0))$ is the solution of an isolated oscillator of the environment when the coupling $u_j = 0$, while $g_j(\tau) = -\theta(\tau)e^{-\eta\tau}\sin(\omega_j\tau)/(m_j\omega_j)$ is the retarded Green's function of the isolated oscillator, see Sec. 10.2.1. The retarded Green's function is 0 when the argument τ is negative, and we have added a small damping parameter $\eta \to 0^+$, so that it decays to 0 when $\tau \to +\infty$. The Green's function satisfies

$$m_j \frac{\partial^2 g_j(t - t')}{\partial t^2} + m_j \omega_j^2 g_j(t - t') = -\delta(t - t').   \tag{9.68}$$

We put this solution of $q_j(t)$ back into the first equation, resulting in

$$M\ddot{X} = -(K + \Delta K)X - \int_{t_0}^{t} \Sigma(t - t')X(t')dt' + R(t),   \tag{9.69}$$

here we define the 'self-energy' as $\Sigma(\tau) = \sum_j u_j^2 g_j(\tau)$, and the 'noise' as $R(t) = \sum_j u_j q_j^0(t)$. The above equation is known as a generalized Langevin equation. To qualify as a stochastic differential equation, we need to define the property of the noise. This is given by saying that at the very beginning when $t_0 \to -\infty$, where the system is decoupled from the environment, the distribution of the environment follows the Gibbs distribution, thus,

$$\langle R(t) \rangle = 0,   \tag{9.70}$$

$$\langle R(t)R(t') \rangle = \sum_j \frac{u_j^2}{\beta m_j \omega_j^2} \cos\big(\omega_j(t - t')\big) = F(t - t').   \tag{9.71}$$

To obtain the noise correlation, we have used the equipartition theorem, $\langle p_j(t_0)p_k(t_0) \rangle = k_B T m_j \delta_{jk}$, $\langle q_j(t_0)q_k(t_0) \rangle = (k_B T/(m_j\omega_j^2))\delta_{jk}$, $\langle p_j(t_0)q_k(t_0) \rangle = 0$, and simplified with trigonometric relations. The noise correlation is closely related to the self-energy $\Sigma(\tau)$. In fact, it is given by a form of the fluctuation-dissipation theorem, best expressed in the frequency domain as [Wang *et al.* (2009)]

$$\tilde{F}(\omega) = -\frac{2}{\beta\omega}\text{Im}\tilde{\Sigma}(\omega).   \tag{9.72}$$

Here the tilde version of F and Σ are the Fourier transforms of the corresponding time domain functions. Finally, to reduce to the Langevin equation, we define $\Sigma(\tau) = d\Gamma(\tau)/d\tau$, more precisely as

$$\Gamma(t) = -\int_{t}^{+\infty} \Sigma(t') \, dt' = \beta F(t).   \tag{9.73}$$

By an integration by parts, we find the ΔK term gets canceled exactly by the $\Gamma(0)$ term, obtaining

$$M\ddot{X} = -KX - \int_{-\infty}^{t} \Gamma(t - t')\dot{X}(t')dt' + R(t). \qquad (9.74)$$

We obtain the standard Langevin equation when it is appropriate to approximate $\Gamma(t - t') \propto \delta(t - t')$ by a delta function, and the noise becomes the white noise. More precisely, we define the spectrum function of the environment as the negative of the imaginary part of the retarded self-energy,

$$J(\omega) = -\text{Im}\tilde{\Sigma}(\omega) = \sum_{j} \frac{\pi u_j^2}{2m_j\omega_j}(\delta(\omega - \omega_j) - \delta(\omega + \omega_j)). \qquad (9.75)$$

The information of the bath is completely encapsulated in this function: the real part of the self energy can be obtained through the Kramers-Kronig relation, and the noise correlation can be expressed as

$$F(\tau) = \int_{-\infty}^{+\infty} \frac{d\omega}{\pi} \frac{J(\omega)}{\beta\omega} e^{-i\omega\tau}. \qquad (9.76)$$

As we can see from Eq. (9.75), if we have only a finite number of oscillators, the spectrum consists of a finite number of δ peaks at frequencies ω_j. However, if we let the number of modes of the oscillators go to infinity, we obtain a smooth function $J(\omega)$. The original Langevin equation is obtained if $J(\omega) = M\gamma\omega$, here γ is the damping parameter related to the viscosity due to Stokes' law introduced earlier.

9.8 Supplementary reading — heat transport in classical phonon systems, a Fokker-Planck equation approach

In this supplementary reading section, we illustrate a non-trivial application of the Fokker-Planck equation to study non-equilibrium steady states and the associated heat transport in the system [Liu et al. (2013)]. The basic idea is that in order to study heat transport we must connect the system of interest to two Langevin baths, call them left and right baths. The baths are at different temperatures so that we can establish a heat flow from hot to cold. The extra degree of freedom of the heat, viewed as a new dynamic variable should be incorporated into a Fokker-Planck equation so that we can compute, not only the heat current, but also the higher order moments, and more generally so-called counting statistics of the heat.

Consider a collection of particles described by the position column vector $x = (x_1, x_2, \ldots, x_N)^T$ and associated conjugate momenta $p = (p_1, p_2, \ldots, p_N)^T$. The superscript T stands for vector or matrix transpose. For simplicity we assume that the masses are all unity, $m_j = 1$; if not, we can always scale the coordinates according to $\sqrt{m_j} x_j$ to make it so. The Langevin equations take the form

$$\dot{x} = p, \tag{9.77}$$

$$\dot{p} = -V'(x) - \Gamma_L p - \Gamma_R p + \xi_L + \xi_R, \tag{9.78}$$

$$\dot{Q} = -p^T \Gamma_L p + p^T \xi_L. \tag{9.79}$$

The first equation says that the velocity is the same as momentum, for each component. This is because we have taken the mass to be 1. The second equation is $F = ma$, where $V(x)$ is the potential energy and the prime denotes gradient with respect to x. The extra terms are viscous damping, random noises, or the baths. Here Γ_L is an $N \times N$ matrix but most of the elements are 0. The nonzero values are located at the upper left corner of the matrix for site 1 or close to it. The nonzero values of entries also correspond to nonzero noises ξ_L through the fluctuation-dissipation theorem:

$$\langle \xi_L(t) \xi_L(t')^T \rangle = 2 \Gamma_L k_B T_L \delta(t - t'). \tag{9.80}$$

The right bath is specified similarly but the matrix Γ_R is nonzero when it is close to the other end of N. The last equation needs some explanation. This equation describes the heat transfer. Q is the amount of heat (or just energy) entering into the system from the left bath, starting from a given time, say $t = 0$. We can see that this is the correct expression for the rate of change by considering the energy of the system, which we define to be $H = \frac{1}{2} p^T p + V$. The energy of the system does not include the damping and random force terms, so it is conserved if it is not connected to the baths. We find that

$$\frac{dH}{dt} = p^T \dot{p} + \dot{x}^T V'(x) = p^T(-\Gamma_L p - \Gamma_R p + \xi_L + \xi_R) = \dot{Q}_L + \dot{Q}_R. \tag{9.81}$$

It is natural to interpret the terms associated with left as from the left bath and right as from the right bath. In steady state, what comes in must go out, $\langle Q_L \rangle + \langle Q_R \rangle = 0$. Thus, it is sufficient to focus on Q_L which we have denoted as just Q. $\dot{Q}$ represents the work done to the system per unit time by the left bath.

Comparing the present set of three coupled equations with the most general form of the Langevin equation, Eq. (9.62), we can identify the defining parameters as

$$X = \begin{pmatrix} x \\ p \\ Q \end{pmatrix}, \quad F(X) = \begin{pmatrix} p \\ -V'(x) - (\Gamma_L + \Gamma_R)p \\ -p^T \Gamma_L p \end{pmatrix}, \quad (9.82)$$

$$g = \begin{pmatrix} 0 & 0 \\ 1 & 1 \\ p^T & 0 \end{pmatrix}, \quad D = \begin{pmatrix} \Gamma_L k_B T_L & 0 \\ 0 & \Gamma_R k_B T_R \end{pmatrix}, \quad \xi = \begin{pmatrix} \xi_L \\ \xi_R \end{pmatrix}.$$

From the general form of Eq. (9.63), we obtain the Fokker-Planck equation for the probability density $\rho(x, p, Q, t)$ of the dynamic variables x, p, and Q, as

$$\frac{\partial \rho}{\partial t} = \left(L^{(0)} + L^{(1)} \left(-\frac{\partial}{\partial Q} \right) + L^{(2)} \left(-\frac{\partial}{\partial Q} \right)^2 \right) \rho. \quad (9.83)$$

Here we split the right-hand side into three terms depending on the order of the partial derivative with respect to Q, where

$$L^{(0)} = -p^T \frac{\partial}{\partial x} + (V')^T \frac{\partial}{\partial p} + \left(\frac{\partial}{\partial p} \right)^T (\Gamma_L + \Gamma_R)p$$

$$+ \left(\frac{\partial}{\partial p} \right)^T (k_B T_L \Gamma_L + k_B T_R \Gamma_R) \frac{\partial}{\partial p}, \quad (9.84)$$

$$-L^{(1)} = p^T \Gamma_L p + k_B T_L \left(p^T \Gamma_L \frac{\partial}{\partial p} + \left(\frac{\partial}{\partial p} \right)^T \Gamma_L p \right), \quad (9.85)$$

$$L^{(2)} = k_B T_L \, p^T \Gamma_L p. \quad (9.86)$$

Here, x and p, and their gradient $\partial/\partial x$ and $\partial/\partial p$ are considered column vectors and the final operators $L^{(i)}$, $i = 0, 1, 2$, are scalars. The partial derivative operator acts on all the factors to the right of it.

We have the following recipe to determine the heat current in steady state. (1) Solve the right eigenvector of the operator $L^{(0)}$ with eigenvalue of zero, i.e., determine ϕ_0 such that $L^{(0)} \phi_0 = 0$. This is the steady state distribution, $\phi_0 = \lim_{t \to \infty} \int \rho(x, p, Q, t) dQ$. (2) Heat current is given by

$$\frac{\langle Q \rangle}{t} = (\psi_0, L^{(1)} \phi_0) = \int L^{(1)} \phi_0 \, dx \, dp. \quad (9.87)$$

Here ψ_0 is the left eigenvector of $L^{(0)}$. As $L^{(0)}$ is not hermitian, the left eigenvector and right eigenvalue are not the hermitian conjugate. It can be shown that the left eigenvector is $\psi_0 = 1$ and the normalization of the right eigenvector is $(\psi_0, \phi_0) = 1$.

We illustrate this procedure with an example of a single site problem first, and then come back to justify Eq. (9.87) later. If the degree of freedom of the problem is one, we have only a single site with coordinate x and momentum p as scalar quantities, then all the vectors and matrices reduce to just scalar numbers. The transpose of vectors is not needed. We can try the following Gibbs distribution for steady state:

$$\phi_0 \propto e^{-\beta H}, \quad H = \frac{p^2}{2} + V(x). \tag{9.88}$$

Substituting this form into the equation $L^{(0)}\phi_0 = 0$, we find that the equation is satisfied if we choose

$$\beta = \frac{\Gamma_L + \Gamma_R}{k_B T_L \Gamma_L + k_B T_R \Gamma_R}, \tag{9.89}$$

i.e., we need to set the overall "temperature" on a weighted average of the left and right bath temperatures based on the couplings. Substituting the solution to the heat current expression, evoking the equipartition theorem for the Gibbs distribution, $\langle p^2 \rangle = 1/\beta$, after some algebra, we find

$$\frac{\langle Q \rangle}{t} = \frac{\Gamma_L \Gamma_R}{\Gamma_L + \Gamma_R} k_B (T_L - T_R). \tag{9.90}$$

This is the prediction of heat transfer per unit time. Note that the result is valid for any form of the potential $V(x)$ and is independent of it. This is a peculiarity of the single site problem.

Another solvable case is when the left bath and right bath are at the same temperature $T_L = T_R = 1/(k_B \beta)$. If we take the Gibbs distribution Eq. (9.88) with any number of degrees of freedom N, we can verify that it is a steady state solution of the Fokker-Planck equation. A non-trivial case is harmonic springs with $V(x) = -Kx$, here K is a positive-definite $N \times N$ force constant matrix. In this case, we seek for an ansatz of the form [Rieder *et al.* (1967)]

$$\phi = \frac{1}{Z} e^{-\frac{1}{2}X^T B^{-1} X}, \quad \int dX \phi = 1. \tag{9.91}$$

Here $X = (x, p)^T$ excludes the Q variable, and B is $2N \times 2N$ symmetric. If this is true, then B is given by the equal-time correlation, $\langle XX^T \rangle$.

We demand that $L^{(0)}\phi = 0$. This requirement is satisfied if

$$AB + BA^T = 2\begin{pmatrix} 0 & 0 \\ 0 & D \end{pmatrix}, \quad A = \begin{pmatrix} 0 & -1 \\ K & \Gamma \end{pmatrix}, \quad (9.92)$$

here $D = k_B T_L \Gamma_L + k_B T_R \Gamma_R$ and $\Gamma = \Gamma_L + \Gamma_R$ are $N \times N$ matrices. We have used the result $\partial\phi/\partial X = -B^{-1}X\phi$, and the steady-state Fokker-Planck equation fixes only the symmetric part of the matrices. Once the steady state is known by solving Eq. (9.92) for B, the current can be obtained again by Eq. (9.87). However, the expression is complicated. Here I just quote the NEGF result, which is

$$\frac{\langle Q \rangle}{t} = \int_0^{+\infty} \frac{d\omega}{2\pi} 4\omega^2 \text{Tr}(G^r \Gamma_L G^a \Gamma_R) k_B(T_L - T_R). \quad (9.93)$$

Here $G^r = (G^a)^\dagger = (\omega^2 I + i\omega\Gamma - K)^{-1}$ is the system retarded Green's function. I leave it as a challenge to the student to show the equivalence of the Fokker-Planck approach and the NEGF expression.

To justify the current formula, Eq. (9.87), we introduce the Fourier transform of the distribution with respect to the variable Q, i.e., we define the characteristic function

$$z(x, p, \lambda, t) = \int_{-\infty}^{+\infty} \rho(x, p, Q, t)e^{\lambda Q} dQ. \quad (9.94)$$

Here λ is a formal parameter. To ensure convergence of the Q integration, we can take λ as a purely imaginary number. Then the inverse transform will have a factor $e^{-\lambda Q}$ and equation for z is obtained with the replacement $-\partial/\partial Q \to \lambda$, i.e.,

$$\frac{\partial z}{\partial t} = \left(L^{(0)} + \lambda L^{(1)} + \lambda^2 L^{(2)} \right) z \equiv Lz. \quad (9.95)$$

We see that $Z = \int dx\, dp\, z(x, p, \lambda, t)$ is the generating function of the energy moments, as $\partial^n Z/\partial\lambda^n|_{\lambda=0} = \langle Q^n \rangle$. It is convenient to study the cumulants

$$\langle\langle Q^n \rangle\rangle = \frac{\partial^n \ln Z}{\partial\lambda^n}, \quad n = 1, 2, \ldots, \quad (9.96)$$

as these quantities are proportional to the time t in the long-time limit. In particular, the first cumulant is the same as the first moment, $\langle\langle Q \rangle\rangle = \langle Q \rangle$. The second cumulant measures the fluctuation, $\langle\langle Q^2 \rangle\rangle = \langle Q^2 \rangle - \langle Q \rangle^2$.

We can solve Eq. (9.95) by expanding the time-dependent solution in terms of the eigenmodes of the operator L. Let $L\varphi = -\mu\varphi$. Since the Langevin system is a dissipative system, we expect all the eigenvalues are

non-negative. After a long time, only the mode with smallest of the eigenvalue survives. The solution takes the form

$$z(x, p, \lambda, t) \approx e^{-\mu t} \varphi(x, p, \lambda), \quad t \to +\infty. \tag{9.97}$$

Here we assume μ is the smallest and associated eigenmode is φ. From
this, we see that the cumulant generating function $\ln Z$ after a long time is
simply $-\mu t$. This shows that all the cumulants are proportional to time t
as claimed. The n-th order culumant is obtained by a Taylor expansion of
μ with respect to λ. We can calculate μ by a perturbation theory according
to quantum mechanics. The current expression, Eq. (9.87), is nothing but
the first order perturbation result for the eigenvalue under the perturbation
of $\lambda L^{(1)}$.

Problems

Problem 9.1. *Quantum particle 'diffuses' differently from classical particles. Consider the discrete Schrödinger equation $i\hbar \frac{d\phi}{dt} = H\phi$ where ϕ
is a column vector of size N and the Hamiltonian is an $N \times N$ hermitian matrix. Specifically we take the nearest-neighbor hopping tight-binding
model with periodic boundary condition, $H_{ij} = -\epsilon$ if $i - j = \pm 1 \pmod{N}$
and 0 otherwise. (a) Solve the time-independent problem, $H\psi = E\psi$ to
find the eigenmodes of the 1D chain. Show the solution is $\psi_j^k = \frac{1}{\sqrt{N}} e^{ikja}$,
$E_k = -2\epsilon \cos(ka)$, here k is the wave vector and a is lattice constant.
(b) We expand the time-dependent solution in terms of eigen modes by
$\Psi(t) = e^{-iHt/\hbar} \Psi(0)$, taking large N limit. We take the initial wave function
to be $\Psi(0)_j = \delta_{0,j}$, i.e., the particle is initially localized at site 0. Show that
the solution can be expressed in terms of the Bessel function of order j as
$\Psi_j(t) = e^{i\pi j/2} J_j\left(\frac{2\epsilon t}{\hbar}\right)$. (c) Show that $\langle x(t) \rangle = 0$ and $\langle x(t)^2 \rangle = 2a^2(\epsilon t/\hbar)^2$,
where $x = na$ is the coordinate of the particle and the average is with respect
to the wave function $\Psi(t)$.*

Problem 9.2. *Assuming a single particle moving according to the
Langevin equation, compute the moments of velocity change in the short-
time limit,*

$$M_n = \lim_{\tau \to 0} \frac{1}{\tau} \left\langle \left(v(\tau) - v\right)^n \right\rangle, \quad v = v(\tau = 0),$$

for $n - 1, 2, \ldots$, and show that $M_n = 0$ if $n > 2$.

Problem 9.3.   *Consider the standard one-dimensional Langevin equation*

$$\frac{dv}{dt} = -\gamma v + \frac{R(t)}{m},$$

where $v(t)$ is velocity, γ is the damping constant, m is mass, and $R(t)$ is the random force with the usual white noise correlation. (a) Find the solution $v(t)$ of the stochastic differential equation in terms of the random force $R(t)$. (b) Compute the velocity random-force correlation, i.e., compute $\langle v(t)R(t')\rangle$ in the long-time limit. Pay attention to the case $t < t'$ and $t > t'$.

Problem 9.4.   *Consider the Langevin equation in the case where the acceleration term mdv/dt is negligible, i.e.,*

$$-k\frac{dx}{dt} + R(t) = 0,$$

with a standard white noise for the random force, i.e., $\langle R(t)\rangle = 0$, $\langle R(t)R(t')\rangle = C\delta(t - t')$. (a) Solve the stochastic differential equation assuming the initial condition $x(0) = 0$; determine the mean-square displacement $\langle x(t)^2\rangle$ as a function of time t. (b) Derive the Fokker-Planck equation for the probability density $\langle P(x,t)\rangle$ for the random variable x.

Problem 9.5.   *A particle moving under gravity in a fluid (sedimentation) or a charged particle in a field follows the Langevin equation with a constant force f, as*

$$m\frac{dv}{dt} = -m\gamma v + f + R(t), \quad dx/dt = v,$$

where $R(t)$ is the random white noise with zero mean and the correlation $\langle R(t)R(t')\rangle = 2m\gamma k_B T\delta(t - t')$. (a) Determine the mobility μ of the particle, which is defined as the proportionality constant of the average velocity to the force, $\langle v\rangle = \mu f$. (b) Derive the associated Fokker-Planck equation from the Langevin equation for the joint probability density $P(v,x,t)$ of velocity v and position x. (c) What is the steady state distribution of the velocity and position, $P(v,x)$?

Problem 9.6.   *Aristotelian physics says that the velocity of a particle is proportional to the force applied to it. We consider such a particle connected to a spring to form an oscillator experiencing a random force (white noise) with the equation*

$$m\gamma\frac{dx}{dt} = -kx + R(t),$$

$$\langle R(t)\rangle = 0, \quad \langle R(t)R(t')\rangle = 2m\gamma k_B T\delta(t - t'),$$

where γ is the damping parameter, m is mass, k is force constant, x is the position of the particle which is a function of time t. The random force $R(t)$ is the standard white noise. (a) Derive a formal solution $x(t)$ expressed in terms of the random force $R(t)$. (b) Derive the associated Fokker-Planck equation for the average probability distribution $\langle P(x,t) \rangle$ of the position variable x. (c) Show that in the long-time limit when equilibrium is reached, the distribution is given by the Gibbs distribution proportional to $e^{-\frac{1}{2}kx^2/(k_BT)}$.

Problem 9.7. *Consider the standard Langevin equation for the velocity, $dv/dt = -\gamma v + R(t)/m$, with $\langle R(t) \rangle = 0$, and $\langle R(t)R(t') \rangle = 2\gamma m k_B T \delta(t - t')$. Here m is particle mass, γ is the damping parameter. (a) Solve the equation formally, assuming steady state is reached, and thus ignoring the transient term to obtain an expression of the velocity $v(t)$ as an integral involving the random noise R. Obtain the position $x(t)$ formally, by integrating the velocity. (b) Determine the correlation function between the noise and the position, i.e., compute $\langle R(t)x(t') \rangle$, assuming steady state so the result should be a function of the time difference, $t - t'$, only. From this result, show explicitly that equal-time correlation $\langle R(t)x(t) \rangle = 0$, justifying Langevin's assumption. (c) Define a new variable $z = d\langle x^2 \rangle/dt$, here the average is over the random noise R. Show that z satisfies the equation $dz/dt + \gamma z = (2k_BT)/m$. Here equipartition theorem is assumed, $m\langle \dot{x}^2 \rangle = k_BT$, for all time t, as well as the conclusion obtained in (b).*

Problem 9.8. *Consider a Langevin equation of the form*

$$\frac{dX}{dt} = -AX + \xi,$$

where the noise satisfies $\langle \xi(t) \rangle = 0$ and $\langle \xi(t)\xi(t')^T \rangle = 2D\delta(t-t')$. Here both A and D are $N \times N$ matrices; X and ξ are column vectors of dimension N. (a) Derive the associated Fokker-Planck equation. (b) Show that a steady state solution of the Fokker-Planck equation is $P(X) \propto e^{-\frac{1}{2}X^T B^{-1} X}$, where the symmetric matrix B satisfies the equation $AB + BA^T = 2D$. Note that this equation implies also $\mathrm{Tr}(A - DB^{-1}) = 0$.

Problem 9.9. *Consider a harmonic oscillator simultaneously experiencing two thermal baths of temperature T_L and T_R modeled by a Langevin equation:*

$$m\frac{dv}{dt} = -kx - m\gamma_L v + \xi_L - m\gamma_R v + \xi_R.$$

The position and velocity are related by $dx/dt = v$. Here m is the mass, k is the force constant, ξ_L and ξ_R are the noises of the left and right bath, which follow the fluctuation-dissipation theorem, relating to the damping coefficients γ_L, γ_R by,

$$\langle \xi_\alpha(t)\xi_\beta(t')\rangle = 2m\gamma_\alpha k_B T_\alpha \delta(t - t')\delta_{\alpha\beta}, \quad \alpha, \beta = L \text{ or } R.$$

k_B is the Boltzmann constant. The first δ is the Dirac delta function and the second is the discrete Kronecker delta. (a) Derive the associated Fokker-Planck equation for the joint probability of velocity v and position x, $P(x, v, t)$. (b) Show that the steady state solution, when the probability distribution becomes independent of the time t at long time, can be written in the Gibbsian form, $P(x, v) = e^{-\beta H}/Z$, here $H = 1/2\,mv^2 + k/2\,x^2$ is the total energy of the oscillator. The choice of Z normalizes the total probability to 1. Determine β in terms of the model parameters given. (c) The average work done by the left bath to the oscillator per unit time is $J = \langle v(-m\gamma_L v + \xi_L)\rangle$. Show that it is given by $J = (\gamma_L \gamma_R)/(\gamma_L + \gamma_R)\,k_B(T_L - T_R)$.

Problem 9.10. *The Kardar-Parisi-Zhang equation is a nonlinear partial differential stochastic equation*

$$\frac{\partial h(x,t)}{\partial t} = \nu\frac{\partial^2 h(x,t)}{\partial x^2} + \frac{\lambda}{2}\left(\frac{\partial h(x,t)}{\partial x}\right)^2 + \xi(x,t),$$

where $\xi(x,t)$ is white gaussian noise with average $\langle\xi(x,t)\rangle = 0$ and correlation $\langle\xi(x,t)\xi(x',t')\rangle = 2D\delta(x - x')\delta(t - t')$. Design a computer simulation algorithm to solve the above equation and verify that the width of height satisfies the Family-Vicsek scaling $W(L, t) \approx L^\alpha f(t/L^z)$ with $\alpha = 1/2$ and $z = 3/2$.

Chapter 10

Systems Near and Far from Equilibrium — Linear Response Theory and Jarzynski Equality

10.1 Introduction

In this chapter, we discuss two extremes, systems that are very close to equilibrium, by a small perturbation, or systems that are truly out of equilibrium. Somehow both cases are closely related to the equilibrium properties. In the first instance, we find that we only need to do the calculation in equilibrium for the correlation functions. This is the linear response theory of Kubo *et al.* [1992]. A slight perturbation actually probes the equilibrium properties related at different space or time points. In the second case, we drive the systems out of equilibrium, here smallness is not assumed. However, surprisingly, certain ensemble averages are related again to the equilibrium free energy difference of the initial and final states. This is the relatively recently discovered Jarzynski equality [Jarzynski (1997)]. This equality generalizes a thermodynamic inequality related to work done to the system and the Helmholtz free energy increase. For this reason, it perhaps does not look too odd that we put these two short topics in the same chapter.

10.2 Linear response theory

We will formulate the theory [Pottier (2010)] in a quantum framework although the theory can also be formulated from classical statistical mechanics. The difference is minor since the von Neumann equation for the density matrix and Liouville's equation for the probability distribution function closely resemble each other. We assume that the system is initially in equilibrium with a quantum-mechanical Hamiltonian H_0 at $t < t_0$. We will use the canonical ensemble so that the equilibrium density matrix takes the form $\rho_{eq} = e^{-\beta H_0}/Z$. Starting from $t > t_0$, we turn on an interaction with the perturbation $V(t)$. This interaction will be explicitly time-dependent, and it will be assumed to be of the form

$$V(t) = -a(t)A, \qquad (10.1)$$

here A is a quantum operator in the Schrödinger picture, and is time independent, while $a(t)$ is a c-number with explicit time dependence. Such form appears, for example, for an atom in an electric field, $-\mathbf{p} \cdot \mathbf{E}(t)$, here the external field is classical, while the dipole moment is proportional to the position operator.

We derive a general formula relating the response to a time-displaced equilibrium average known as the retarded Green's function. To achieve this, we need to determine how the density matrix evolves in time. This is the solution of the von Neumann equation, $i\hbar d\rho/dt = [H_0 + V(t), \rho]$. It is convenient to work in the interaction picture which removes the H_0 term, instead of the usual Schrödinger picture. Both the density matrix and operators are transformed in the same way by a unitary transform:

$$\rho_I(t) = e^{iH_0 t/\hbar}\rho(t)e^{-iH_0 t/\hbar}, \qquad (10.2)$$

$$A_I(t) = e^{iH_0 t/\hbar}A\,e^{-iH_0 t/\hbar}, \qquad (10.3)$$

here the operators without a subscript are in the Schrödinger picture. The transformation keeps the trace invariant. Differentiating the interaction picture density matrix $\rho_I(t)$ with respect to time and using the original von Neumann equation, we find that we can cancel out the H_0 term, obtaining

$$i\hbar\frac{d\rho_I(t)}{dt} = [V_I(t), \rho_I(t)], \qquad (10.4)$$

here the interaction picture interaction term is $V_I(t) = e^{iH_0 t/\hbar}V(t)e^{-iH_0 t/\hbar}$. This equation can be solved iteratively. First, we integrate from t_0 to t on both sides of the equation, obtaining

$$\rho_I(t) = \rho_{eq} + \frac{1}{i\hbar}\int_{t_0}^{t} [V_I(t'), \rho_I(t')]\,dt'. \qquad (10.5)$$

In fact this is still an equation for $\rho_I(t)$ as the function appears on both the left and right sides. However, we can use the result obtained so far and substitute the expression back to itself repeatedly, getting

$$\rho_I(t) = \rho_{\text{eq}} + \frac{1}{i\hbar} \int_{t_0}^{t} \left[V_I(t'), \rho_{\text{eq}}\right] dt'$$

$$+\frac{1}{(i\hbar)^2} \int_{t_0}^{t} dt_1 \int_{t_0}^{t_1} dt_2 \left[V_I(t_1), [V_I(t_2), \rho_{\text{eq}}]\right] + \cdots . \quad (10.6)$$

This result can be formally expressed as $T e^{\frac{1}{i\hbar} \int_{t_0}^{t} [V(t'), \cdot] dt'} \rho_{\text{eq}}$, where T is the time order super-operator. For a linear response calculation, it is sufficient we stop at the first-order term in V.

Having obtained an approximate result for the density matrix evolving in time, we consider the expectation value of another dynamic variable, call it B. The interaction picture quantity is $B_I(t)$, which has the same transformation, as in Eq. (10.3). Since the expectation value is invariant with respect to a unitary transform, we find

$$\langle B(t) \rangle = \text{Tr}\big(\rho_I(t) B_I(t)\big)$$

$$= \text{Tr}(\rho_{\text{eq}} B) - \frac{i}{\hbar} \int_{t_0}^{t} \text{Tr}\big([V_I(t'), \rho_{\text{eq}}] B_I(t)\big) dt'$$

$$= \frac{i}{\hbar} \int_{t_0}^{t} \text{Tr}\big(\rho_{\text{eq}}[B_I(t), A_I(t')]\big) a(t') dt'. \quad (10.7)$$

In going over to the last line above, we assume that the variable B is such that its equilibrium average $\text{Tr}(\rho_{\text{eq}} B)$ is 0. This is not really a restriction, for if the average is not 0, we can define a new observable as $B - \langle B \rangle$. Also, we have expanded the commutator and used the cyclic permutation property of trace to obtain the last form. Finally, we define the retarded Green's function with two variable B and A as

$$G_{BA}(t, t') = -\frac{i}{\hbar}\theta(t - t')\text{Tr}\big(\rho_{\text{eq}}[B_I(t), A_I(t')]\big). \quad (10.8)$$

The Heaviside step function $\theta(t)$ is defined as $\theta(t) = 1$ if $t > 0$ and 0 if $t < 0$ (exactly what value to define at $t = 0$ is debatable). Then the linear response result is

$$\langle B(t) \rangle = -\int_{t_0}^{t} G_{BA}(t, t') a(t') \, dt'. \quad (10.9)$$

$\chi(t, t') = -G(t, t')$ is also known as a generalized susceptibility. If we take $t_0 \to -\infty$, we can Fourier transform them into the frequency domain. In particular, we can show that the function of $G_{BA}(t, t') = G_{BA}(t - t')$ is in fact a function of the time difference only, because systems in equilibrium have time translational invariance. Thus, $G_{BA}(t)$ is Fourier transformed to $\tilde{G}(\omega)$ of a single argument. In the frequency domain, applying the convolution theorem, the linear response result is simply

$$\tilde{B}(\omega) = -\tilde{G}(\omega)\tilde{a}(\omega). \tag{10.10}$$

The Fourier transforms are defined in the same way as by Eq. (9.37).

The commutator defining the retarded Green's function does not have a nice classical analogue — the commutator becomes 0 if replaced by classical numbers, but the denominator also has an $\hbar$ which should approach 0 as well. To overcome this difficulty, Kubo introduced another correlation, known as the Kubo correlation, defined for equilibrium average as, for any quantum operators a and b by

$$\langle a;b \rangle = \frac{1}{\beta}\int_0^\beta d\lambda\, \mathrm{Tr}\big(\rho_{\mathrm{eq}}\, e^{\lambda H_0} a\, e^{-\lambda H_0} b\big), \quad \beta = \frac{1}{k_B T}. \tag{10.11}$$

We have $\langle a;b \rangle = \langle b;a \rangle$, although a and b may be non-commuting. We can prove the following Kubo identity relating the commutator with the Kubo correlator, as

$$\beta\langle \dot{a};b \rangle = \frac{1}{i\hbar}\mathrm{Tr}\big(\rho_{\mathrm{eq}}[a,b]\big). \tag{10.12}$$

Here $\dot{a} = \frac{1}{i\hbar}[a,H_0]$ is from the Heisenberg equation of motion. As a result of these transformations, we can also write

$$\langle B(t) \rangle = -\beta \int_{t_0}^{t} \langle \dot{B}_I(t); A_I(t') \rangle a(t')\, dt'. \tag{10.13}$$

This equation is then applicable to classical systems. We simply replace the quantum operators by numbers and replace the quantum average by the corresponding classical ensemble average (at different times).

10.2.1 *Example: a quantum oscillator under external driven force*

As a simple application of the above result, let us consider a simple quantum harmonic oscillator with the Hamiltonian

$$H_0 = \frac{p^2}{2m} + \frac{1}{2}m\Omega^2 x^2, \tag{10.14}$$

here m is mass and Ω is the oscillator angular frequency, the momentum and coordinate satisfy the canonical commutation relation $[x,p] - i\hbar$. It is more convenient if we transform into the ladder operators a and $a^\dagger$ with commutator $[a, a^\dagger] = 1$. Then the relations between the old and new operators are

$$x = \sqrt{\frac{\hbar}{2m\Omega}}(a + a^\dagger), \quad p = -im\Omega\sqrt{\frac{\hbar}{2m\Omega}}(a - a^\dagger). \tag{10.15}$$

Using the creation and annihilation operators, we can write the Hamiltonian as $H_0 = \hbar\Omega(a^\dagger a + 1/2)$, and the Heisenberg equation for a is particularly simple,

$$\frac{da}{dt} = \frac{1}{i\hbar}[a, H_0] = -i\Omega\, a. \tag{10.16}$$

Its solution is $a(t) = a_I(t) = a\,e^{-i\Omega t}$. Note that in the previous section we call this time-dependent Heisenberg operator the interaction picture operator with a subscript I.

Assuming at $t_0 \to -\infty$ the oscillator is in thermal equilibrium with a canonical distribution, we perturb the system with a time-dependent force, so that $V(t) = -xf(t)$ (remember the negative gradient with respect to the coordinate is the force). We ask what the average position at a later time is for the oscillator. According to the linear response theory, this quantity is given by

$$\langle x(t) \rangle = -\int_{-\infty}^{+\infty} G_{xx}(t - t')f(t')dt', \tag{10.17}$$

where the retarded coordinate Green's function is

$$G_{xx}(t) = -\frac{i}{\hbar}\theta(t)\langle [x(t), x(0)] \rangle_0, \tag{10.18}$$

here the average $\langle \cdots \rangle_0$ is evaluated in the canonical ensemble defined by H_0. Because of the time-translational invariance, we can always choose the second time $t' = 0$. Using the explicit expression in Eq. (10.15) together with the solution of the Heisenberg equation, the commutator can be calculated, $[x(t), x(0)] = \frac{\hbar}{2m\Omega}[e^{-i\Omega t}a + e^{i\Omega t}a^\dagger, a + a^\dagger] = \frac{\hbar}{2m\Omega}(e^{-i\Omega t} - e^{i\Omega t})$. The retarded Green's function is easily evaluated as

$$G_{xx}(t) = -\theta(t)\frac{\sin(\Omega t)}{m\Omega}. \tag{10.19}$$

A peculiar feature of this oscillator example is that the result does not depend on the initial distribution. Any density matrix ρ at t_0 will produce the same result. It is not so for more general interacting systems. To go into

Fourier space, we need to add a damping factor $e^{-\eta t}$ as the sin function does not have a well-defined Fourier transform. With this damping, the Green's function in the frequency domain is

$$\tilde{G}(\omega) = \frac{1}{m\big((\omega + i\eta)^2 - \Omega^2\big)}, \quad \eta \to 0^+. \tag{10.20}$$

A single mode of phonon or photon is a quantum oscillator. Thus, the Green's function we obtained here will find good utility in the many-body theory of phonons or photons.

10.3 Kubo-Greenwood formula

In this section, we continue with the application of the Kubo linear response theory to calculate a model electronic conductivity. We assume the electrons are nearly free with a long relaxation time to describe their dissipation or damping by a self energy in the electron Green's function. We recall Ohm's law

$$j_\alpha(\omega) = \sum_{\beta = x, y, z} \sigma_{\alpha\beta}(\omega) E_\beta(\omega). \tag{10.21}$$

Here the current $\mathbf{j}$, electric field $\mathbf{E}$, and conductivity tensor σ are frequency resolved and the Greek letters α and β denote the Cartesian directions. We define the Fourier transform in frequency according to, e.g.,

$$\tilde{j}_\alpha(\omega) = \int_{-\infty}^{+\infty} j_\alpha(t) e^{i\omega t} dt. \tag{10.22}$$

With somewhat of an abuse of notation, we drop the tilde to the Fourier transform of time-dependent quantities. Our task is to give an expression for the alternating current conductivity in terms of the materials properties. Here we'll assume that our material is a metal or semi-conductor consisting of solely electrons described by a tight-binding Hamiltonian $\hat{H} = \sum_{i,j} c_i^\dagger H_{ij} c_j$. Here c_j and $c_i^\dagger$ are the second quantized Fermi annihilation operator and creation operator at site j and i. The matrix H is hermitian, i.e., $H_{ji}^* = H_{ij}$. We can choose a gauge such that the scalar potential ϕ is 0 and electric field is related to the vector potential by

$$\mathbf{E} = -\frac{\partial \mathbf{A}}{\partial t} \quad \to \quad E_\beta(\omega) = i\omega A_\beta(\omega). \tag{10.23}$$

The electrons are coupled to the field by $H' = -\int \mathbf{j} \cdot \mathbf{A} \, d^3\mathbf{r}$ which we treat as a perturbation to the electrons.

In the next step, we need an explicit expression of the electric current in terms of electron degrees of freedom. Recall the Hamiltonian of a single particle in a vector potential is obtained by replacing the momentum operator $\mathbf{p}$ by $\mathbf{p} + e\mathbf{A}$, here electron charge is $-e$. This produces two terms in the current, a diamagnetic term proportional to $\mathbf{A}$ and a term proportional to the conjugate momentum. For the moment, we'll ignore the diamagnetic term which has a simple form proportional to the electron density, and focus only on the nontrivial paramagnetic term.

It is convenient to consider the "total" current by integration over a large volume

$$\mathbf{J} = \int_V \mathbf{j}\, d^3\mathbf{r}, \tag{10.24}$$

which is the $\mathbf{k} \to 0$ limit of the space Fourier transform of the current. The total current is obtained by the continuity equation of charge conservation, $\dot{\rho} + \nabla \cdot \mathbf{j} = 0$. In $\mathbf{k}$ space this is $\dot{\rho} + i\mathbf{k} \cdot \mathbf{j_k} = 0$. Here ρ is the charge density

$$\rho(\mathbf{r}, t) = (-e) \sum_j c_j^\dagger c_j \delta(\mathbf{r} - \mathbf{R}_j). \tag{10.25}$$

Here $\mathbf{R}_j$ is the position of the site j. The time derivative of the density is obtained from the Heisenberg equation of motion, $\dot{\rho} = \frac{1}{i\hbar}[\rho, \hat{H}]$. Carrying out the commutation relation calculation, taking into account that c_j and $c_i^\dagger$ are fermion operators, and taking the zero wave vector $\mathbf{k}$ limit, we obtain a formula for the total current as

$$\mathbf{J} = (-e) \sum_{ij} c_i^\dagger \mathbf{V}_{ij} c_j, \qquad V_{jk}^\alpha = \frac{1}{i\hbar} H_{jk}(R_j^\alpha - R_k^\alpha). \tag{10.26}$$

Here we define the "velocity" operator matrix element according to the second formula. The index α takes x, y, z, specifying the three Cartesian directions.

If we have a lattice periodic system (crystal) with one degree of freedom per unit cell, we can perform a Fourier transform for c_j into reciprocal space. Then we will find that both the Hamiltonian and the current will be diagonal in $\mathbf{k}$ space. As a result, the velocity matrix is then nothing but the group velocity $\partial \epsilon_\mathbf{k}/\partial(\hbar\mathbf{k})$ of the electron. Here $\epsilon_\mathbf{k}$ is the dispersion relation of the electron with $\mathbf{k}$ varying in the first Brillouin zone. For generality of the formulation, here we will not assume lattice periodicity, thus the result can be applied to disordered systems or nanostructures.

With the definition of the total current, if we apply a spatially uniform electric field (and spatially uniform vector potential), the perturbation term can be written as

$$H' = -\sum_\beta J_\beta A_\beta(t). \tag{10.27}$$

Comparing the general formalism of the Green-Kubo formula, Eq. (10.1), with the perturbation and the resulting response of Eq. (10.10), we can identify $A_\beta(t)$ as the scalar time-dependent parameter $a(t)$, the operator $\mathbf{J}$ as A, as well as response B. We have

$$\langle J_\alpha(\omega)\rangle = -\sum_\beta G_{J_\alpha J_\beta}(\omega)A_\beta(\omega)$$

$$= -\sum_\beta \frac{1}{i\omega}G_{J_\alpha J_\beta}(\omega)E_\beta(\omega) = V\sum_\beta \sigma_{\alpha\beta}(\omega)E_\beta(\omega). \tag{10.28}$$

Thus, we see that the frequency-dependent dynamic conductivity is given by the current-current retarded Green's function divided by $-i\omega$. We need a volume factor here since $\mathbf{J} = \mathbf{j}V$, and the conductivity is defined in terms of the current density $\mathbf{j}$. We can write out the conductivity in terms of the retarded Green's function explicitly as [Mahan (2000)]

$$\sigma_{\alpha\beta}(\omega) = \frac{1}{\hbar\omega V}\int_0^\infty \langle [J_\alpha(t), J_\beta(0)]\rangle e^{i\omega t-\eta t}dt. \tag{10.29}$$

Here the time-dependence of the total current is due to Heisenberg evolution of the operator. The square brackets denote the commutator. The angular brackets denote thermodynamic average in an equilibrium grand-canonical ensemble with the density matrix $\hat\rho \propto e^{-(\hat H-\mu\hat N)/(k_B T)}$. We make the Fourier transform by the time integration. The retarded Green's function is 0 when $t < 0$ so the integral runs from 0 to positive infinity only. We add a small damping term $\eta > 0$. This guarantees convergence of the integration. More importantly, this adds a small damping to the electrons so that the electrons acquire a finite relaxation time with a finite conductivity.

The above result is the formal Green-Kubo formula for electric conductivity. To obtain a more concrete result for the non-interacting electron model, we focus on the current-current correlation function and apply Wick's theorem to express them in terms of the lesser and greater Green's functions of the electrons. They are defined by

$$G_{jk}^>(t) = -\frac{i}{\hbar}\langle c_j(t)c_k^\dagger(0)\rangle, \quad G_{jk}^<(t) = \frac{i}{\hbar}\langle c_k^\dagger(0)c_j(t)\rangle. \tag{10.30}$$

The retarded Green's function of the electron is defined by $G^r(t) = \theta(t)$ $\big(G^>(t) - G^<(t)\big)$. The commutator of the current gives $\langle[J_\alpha(t), J_\beta(0)]\rangle =$ $\langle J_\alpha(t)J_\beta(0)\rangle - \langle J_\beta(0)J_\alpha(t)\rangle$. Focusing on the first term, and using the current expression, $J_\alpha = (-e)\sum_{ij} c_i^\dagger V_{ij}^\alpha c_j$, we obtain expectation values of four fermion operators of the form

$$\langle c_i^\dagger(t)c_j(t)c_k^\dagger(0)c_l(0)\rangle. \tag{10.31}$$

Wick's theorem says that we can break this up as a sum of products of all possible pairs. Since cc or $c^\dagger c^\dagger$ pairs are zero, as they do not conserve particle number, we only have creation operators paired with annihilation operators or vice versa. Two special points are worth noting for fermion operators. (1) We must maintain the original order, i.e., if site i is before site j, this order need to be maintained in the pairing. (2) To bring two operators next to each other, if it requires an odd number of passes through other fermion operators, we need to have a minus sign. This concludes the statement of Wick's theorem. A formal proof of its validity is a bit too long to state here.

Using this rule, Eq. (10.31) could have two terms but the equal-time term is canceled by the reverse order term in the commutator. The non-vanishing term is then given by

$$\langle c_i^\dagger(t)c_l(0)\rangle\langle c_j(t)c_k^\dagger(0)\rangle = (-i\hbar)G_{li}^<(-t)(i\hbar)G_{jk}^>(t). \tag{10.32}$$

Putting all the terms together, we can express the average value of the current-current commutator as

$$\langle[J_\alpha(t), J_\beta(0)]\rangle = e^2\hbar^2\Big(\text{Tr}\big(G^<(-t)V^\alpha G^>(t)V^\beta\big)$$

$$- \text{Tr}\big(G^>(-t)V^\alpha G^<(t)V^\beta\big)\Big). \tag{10.33}$$

Here $G^<$, $G^>$, and V^α are viewed as matrices indexed by the electron sites, and the trace is over the site variables. To obtain the final expression in the frequency domain, we will quote the so-called "fluctuation-dissipation" theorem in the frequency domain [Haug and Jauho (1996)]:

$$G^<(\omega) = -f(\omega)\big(G^r(\omega) - G^a(\omega)\big), \tag{10.34}$$

$$G^>(\omega) = \big(1 - f(\omega)\big)\big(G^r(\omega) - G^a(\omega)\big). \tag{10.35}$$

Here $f(\omega) = 1/\big(e^{(\hbar\omega-\mu)/(k_B T)} + 1\big)$ is the Fermi function, G^r is the retarded Green's function, and $G^a = (G^r)^\dagger$ is the advanced Green's function. We will define $A(\omega) = i(G^r - G^a)$ to be the spectrum function. Expressing all the time-dependence by their corresponding Fourier transformed ones, and integrating over the time t, we finally obtain

$$\sigma_{\alpha\beta}(\omega) = \frac{ie^2\hbar^2}{\hbar\omega V}\int\frac{d\omega'}{2\pi}\int\frac{d\omega''}{2\pi}\frac{f(\omega')-f(\omega'')}{\omega+\omega'-\omega''+i\eta}\mathrm{Tr}\big(A(\omega')V^{\alpha}A(\omega'')V^{\beta}\big).$$

The DC conductivity is the real part of the above function when ω goes to 0. Using the Plemelj formula to separate out the real part containing the δ function,

$$\frac{1}{x+i\eta} = P\frac{1}{x} - i\pi\delta(x), \quad \eta \to 0^+, \tag{10.36}$$

we obtain

$$\sigma_{\alpha\beta} = \lim_{\omega\to 0}\mathrm{Re}\,\sigma_{\alpha\beta}(\omega)$$

$$= \frac{e^2\hbar^2}{2V}\int_{-\infty}^{+\infty}\frac{dE}{2\pi\hbar}\left(-\frac{\partial f}{\partial E}\right)\mathrm{Tr}\big(A(E)V^{\alpha}A(E)V^{\beta}\big). \tag{10.37}$$

Here we have made a change of variable, $E = \hbar\omega$. This is the Kubo-Greenwood formula for electron conductivity. Note that in our derivation, we assumed spinless electrons. We can easily incorporate the spin degrees in the trace, i.e., interpreting the trace as sum over both site and spin, (j, σ).

We briefly comment on the omitted diamagnetic term. This term gives an extra current $\mathbf{j} = -e^2 n\mathbf{A}/m$ where n is the number of electrons per unit volume. Expressing this result in terms of electric field, we see that it corresponds to a purely imaginary contribution to the frequency dependent conductivity, $ie^2 n/(m\omega)\delta_{\alpha\beta}$. Thus, this term does not contribute to the dissipative part of the conductivity.

10.4 Jarzynski equality

We discuss first the thermodynamic inequality. Fundamental to irreversible thermodynamics is the Clausius inequality, $TdS \geq \delta Q$, which states for an adiabatic process with $\delta Q = 0$, entropy cannot decrease. Using the first law of thermodynamics, we have $dU = \delta Q + \delta W$. Thus, we also have

$$TdS \geq dU - \delta W. \tag{10.38}$$

Consider an isothermal process so that temperature is a constant. With the definition of the Helmholtz free energy as $F = U - TS$, and $dT = 0$, we can also express the above inequality as

$$\delta W \geq dF, \quad T = \text{const.} \tag{10.39}$$

This inequality has the following interpretation: the amount of work done to the system must be always larger than or at least equal (for a reversible process) to the free energy increase. We can also write the inequality in an integral form, $W \geq F_B - F_A = \Delta F$, here A denotes the initial state and B the final state. W is the work done to the system by going from state A to state B, and ΔF is the free energy increase. Jarzynski turns this inequality into an equality, as

$$\langle e^{-\beta w} \rangle = e^{-\beta \Delta F}, \quad \beta = \frac{1}{k_B T}. \tag{10.40}$$

Here the angular brackets denote statistical average, or more precisely, an average over the canonical distribution of the initial equilibrium state. From a statistical-mechanical point of view, the microscopic initial states are not a single state but a distribution, thus the work done going from macroscopic state A to B has many possibilities and the work done w is a fluctuational quantity. Its average in the form of an exponential is related to the free energy difference. We will define w more precisely below. Here we note that the Jarzynski equality is consistent with the thermodynamic inequality. Since the exponential function is a convex function, we have $\langle e^x \rangle \geq e^{\langle x \rangle}$. Using the convexity property of the exponential function, we find $e^{-\beta \langle w \rangle} \leq e^{-\beta \Delta F}$ or $\langle w \rangle \geq \Delta F$. Here we can identify the average work with the thermodynamic work W.

To prove the Jarzynski equality, we need to define precisely how we are going to move the system from initial state A to final state B. Here we treat the system classically, and assume that the system is defined by a Hamiltonian with a parameter, $H(\lambda)$. The parameter λ will be a function of time, $\lambda(t)$, with $\lambda_A = \lambda(0)$ for the initial state A and $\lambda_B = \lambda(t_B)$ for the final state at time $t_B > 0$. The choice of this function is arbitrary but fixed. For example, this $\lambda(t)$ may represent pulling of the molecules in a definite way. We may repeat the experiments many times, but the pulling is fixed. We call the function $\lambda(t)$ a protocol. The associated Hamiltonians are $H_A = H(\lambda_A)$ and $H_B = H(\lambda_B)$, respectively. Initially at time $t \leq 0$ we assume that the system is in thermal equilibrium with the distribution $\rho(\Gamma, 0) \propto e^{-\beta H_A(\Gamma)}$, here $\Gamma = (q_1, q_2, \ldots, p_1, p_2, \ldots)$ is the set of coordinates and conjugate momenta in phase space.

A crucial ingredient in proving the Jarzynski equality is the definition of the microscopic work. We use

$$w = \int_0^{t_B} dt \, \dot{\lambda} \frac{\partial H}{\partial \lambda} = \int_{\lambda_A}^{\lambda_B} d\lambda \frac{\partial H}{\partial \lambda} = H(\lambda_B, \Gamma) - H(\lambda_A, \Gamma_0). \tag{10.41}$$

This definition of work is known as 'inclusive' work [Campisi *et al.* (1997)], as oppose to another definition by the force times displacement. We also note that w depends on the initial microscopic state Γ_0, while the final state $\Gamma = \Gamma_{t_B}$ is fixed by the equation of motion of the Hamiltonian system. We could give a formal expression relating the two as $\Gamma = T \exp\left(\int_0^{t_B} dt(\,\cdot\,, H_t)\right)\Gamma_0$, here T is the time order operator, and $(\,\cdot\,, H_t)$ is the Poisson bracket.

Our aim is to compute the canonical average

$$\left\langle e^{-\beta w} \right\rangle = \int d\Gamma_0 \, \rho(\Gamma_0, 0) e^{-\beta w(\Gamma_0)}. \tag{10.42}$$

This seems to be very difficult as we do not know the final state Γ unless we actually solve the Hamilton equations of motion. The trick to move forward is to make a change of variable from Γ_0 to Γ. Since such a change of variable is a canonical transform, the Jacobian is 1 (show that this is so even though the Hamiltonian is time-dependent). With $\rho(\Gamma_0, 0) e^{-\beta w} = \frac{1}{Z_A} e^{-\beta H_A} e^{-\beta(H_B - H_A)} = e^{-\beta H_B(\Gamma)}/Z_A$ and a change of variable from Γ_0 to Γ, we find

$$\left\langle e^{-\beta w} \right\rangle = \frac{1}{Z_A} \int d\Gamma e^{-\beta H_B(\Gamma)} = \frac{Z_B}{Z_A} = e^{-\beta(F_B - F_A)}. \tag{10.43}$$

This is the Jarzynski equality.

It is clear the final state at time t_B is not distributed according to the canonical distribution with the Hamiltonian H_B. This is because the system, due to the explicit time-dependent drive, is out of equilibrium. Thus, the meaning of the free energy F_B can be interpreted as the equilibrium free energy when the system is re-attached to a thermal bath and equilibrium is re-established after a long time after t_B. The Jarzynski equality has some generality. For example, the same result holds if we consider a Brownian particle within a Langevin or Fokker-Planck framework. It falls into a general consideration of so-called fluctuation theorems. As an example of such types of relations, we consider the following thought experiment. Consider two insulating solids connected in some way by a bridge. The solids are maintained at temperature T_L and T_R which are different. As a result, heat conduction occurs. Let us assume that the bridge is initially disconnected so there is no heat conduction at $t < 0$. We then suddenly connect them for a duration of time t_B and disconnect them again. We ask the following question: what is the heat Q transferred from the left to right during this process? Due to the microscopic nature, the answer is not a fixed number but a probability distribution. Let $P(Q)dQ$ be the probability for the heat transferred to be in the range Q and $Q+dQ$, it may be very difficult to obtain this probability distribution, but the following ratio can be established [Agarwalla *et al.* (2012)]:

$$\frac{P(Q)}{P(-Q)} = e^{Q(\beta_R - \beta_L)}, \quad \beta_{L,R} = \frac{1}{k_B T_{L,R}}. \tag{10.44}$$

This is known as the Gallavatti-Cohen symmetry. Can heat then flow from cold to hot? Based on this equation, if $T_L < T_R$ (then $\beta_R - \beta_L < 0$), $Q > 0$ is possible but with exponentially small probability. Heat flowing from hot to cold has a much higher probability.

Problems

Problem 10.1. *Show that the Jacobian of* Γ_0 *to* Γ, *i.e., the determinant formed by the partial derivatives,* $\partial \Gamma_i / \partial (\Gamma_0)_j$, *is 1 for a general Hamiltonian system where the Hamiltonian may explicitly depend on time. Here* Γ_0 *is the initial state at time* $t = 0$ *and* Γ *is the phase point at a later (fixed) time* t.

Problem 10.2. *Based on the definition of the Kubo correlation,* $\langle a; b \rangle = \frac{1}{\beta} \int_0^\beta d\lambda \, \mathrm{Tr}(\rho_{\mathrm{eq}} e^{\lambda H} a \, e^{-\lambda H} b)$, *and* $\rho_{\mathrm{eq}} = e^{-\beta H}/Z$, *show that for finite quantities* a *and* b,

$$\langle a; b \rangle = \langle b; a \rangle, \quad \langle a; a \rangle \geq 0, \quad \langle \dot{a}; b \rangle = -\langle a; \dot{b} \rangle.$$

Problem 10.3. *Prove the Kubo identity*

$$\beta \langle \dot{a}; b \rangle = \frac{1}{i\hbar} \mathrm{Tr}(\rho_{\mathrm{eq}}[a, b]).$$

Here $\dot{a} = \frac{1}{i\hbar}[a, H]$ *is from the Heisenberg equation of motion,* $\rho_{\mathrm{eq}} = e^{-\beta H}/Z$ *is the canonical distribution,* $\beta = 1/(k_B T)$.

Problem 10.4. *Heat conduction in a solid is caused by inhomogeneity in temperature, thus we can describe the situation by a density operator of a canonical distribution as* $\rho \propto \exp\left(-\int d^3\mathbf{r}\, \beta(\mathbf{r})\epsilon(\mathbf{r})\right)$, *where* $\epsilon(\mathbf{r})$ *is energy density per unit volume.* (a) *Expanding around an equilibrium value,* $\beta(\mathbf{r}) = \beta + \delta\beta(\mathbf{r})$, *show that the perturbation Hamiltonian is* $V' = -\frac{1}{T} \int d^3\mathbf{r}\, \delta T(\mathbf{r})\epsilon(\mathbf{r})$ *and the unperturbed Hamiltonian is* $H_0 = \int d^3\mathbf{r}\, \epsilon(\mathbf{r})$. *Unfortunately such consideration is wrong for the perturbation by a minus sign. The correct answer, due to Luttinger, is the one without the extra minus sign,* $V = -V'$. (b) *Using* V *as a perturbation, applying the Kubo linear response theory, and noticing continuity equation relating the energy* density with energy current, $\partial \epsilon/\partial t + \nabla \cdot \mathbf{j} = 0$, derive the Kubo formula for heat conductivity,

$$\kappa_{\alpha\beta} = \frac{1}{k_B T^2 V} \int_0^\infty dt\, e^{-\eta t} \langle J_\beta(0); J_\alpha(t)\rangle,$$

where the Fourier law is given by $j_\alpha = -\sum_\beta \kappa_{\alpha\beta}\partial T/\partial x^\beta$, the angular bracket $\langle\ ;\ \rangle$ denotes the Kubo correlation, and $J_\alpha = \int d^3\mathbf{r}\, j_\alpha$ is the total current. The index α or β takes the Cartesian components.

Problem 10.5. *Show that on a translational invariant Bravais lattice with each site occupied by one electron state, the velocity-vector matrix of $\dot{\mathbf{r}} = \mathbf{v} = [\mathbf{r}, H]/(i\hbar)$ in a basis that the position operator is diagonal, $\mathbf{V}_{jk} = \frac{1}{i\hbar} H_{jk}(\mathbf{R}_j - \mathbf{R}_k)$, when Fourier transformed into the reciprocal space,*

$$\mathbf{v}_{\mathbf{k}} = \sum_j \mathbf{V}_{jk} e^{-i\mathbf{k}\cdot(\mathbf{R}_j - \mathbf{R}_k)},$$

is given by the group velocity of the electron band, where the dispersion relation $\epsilon_{\mathbf{k}}$ is similarly the Fourier transform of the Hamiltonian H_{jk} from real space to reciprocal space.

Problem 10.6. *Based on the definition of Green's functions for electrons, show that (a) $G^> - G^< = G^r - G^a$. Here the identity is valid as a matrix equation both in real time t or in energy domain $E = \hbar\omega$. (b) Prove the fluctuation-dissipation relation, $G^< = -f(G^r - G^a)$, defined in energy domain. Here $f = 1/(e^{\beta(E-\mu)}-1)$ is the Fermi function. Hint: expand both sides of the equation in the eigenstates of the Hamiltonian, $\hat{H}|n\rangle = E_n|n\rangle$, and using the grand-canonical distribution $\langle \cdots \rangle = \mathrm{Tr}\,(e^{-\beta(\hat{H}-\mu\hat{N})} \cdots)/\Xi = \sum_n \frac{1}{\Xi} e^{-\beta(E_n - \mu N_n)}|n\rangle\langle n| \cdots$.*

Problem 10.7. *Consider a classical particle of mass m trapped in a parabolic potential in three dimensions, $V(\mathbf{r}) = 1/2\, a\, r^2$, here a is some control constant, and $\mathbf{r} = (x, y, z)$ is a three-dimensional position vector. The Hamiltonian of the particle is $H = p^2/(2m) + V(\mathbf{r})$, where $\mathbf{p}$ is the momentum vector. (a) Compute the partition function Z of the single particle system in the canonical ensemble. From it, determine the Helmholtz free energy F. (b) The particle is assumed initially in thermal equilibrium at temperature $T = 1/(k_B\beta)$. If we drive the system by changing the shape a of the confining potential, from a_i to the final a_f, in some way $a(t)$, what is the expectation value of $e^{-\beta w}$ according to the Jarzynski equality for the nonequilibrium process? Give the definition of the work w.*

Problem 10.8. *The Jarzynski equality relates the equilibrium free energy to a nonequilibrium process. Consider a classical Hamiltonian of the form $H(t) = p^2/(2m) + 1/2\,k\big(x - vt\theta(t)\big)^2$, which represents a harmonic oscillator with the center of the force moving with velocity v when $t > 0$. Here the step function $\theta(t) = 0$ if $t < 0$ and 1 when $t \geq 0$. (a) Determine the free energy difference $F(t) - F(0)$ associated with the Hamiltonian $H(t)$ and $H(0)$. Here t is a fixed time. (b) Solve the equation of motion of the oscillator to determine $x(t)$ and $p(t)$ in terms of the initial values of position x_0 and momentum p_0, as well as time t. (c) Give the expression for the work W done going from microscopic state (x_0, p_0) to the state $(x(t), p(t))$, according to Jarzynski. Express the work W as a function of the initial state only. (d) Assuming that the initial distribution of (x_0, p_0) follows the canonical distribution determined by the initial Hamiltonian $H(0)$, show that the exponential average of the work, $\langle e^{-\beta W}\rangle_0$, by explicit calculation of the average, is given by the Jarzynski equality. Here $\beta = 1/(k_B T)$, k_B is the Boltzmann constant and T is temperature.*

Chapter 11

The Boltzmann Equation

11.1 Introduction

In this chapter, we present the Boltzmann equation [Huang (1987); Smith and Jensen (1989); Harris (2004)]. In 1872, Boltzmann put down his famous equation and also presented a formula for entropy in terms of the distribution function, f. This is, for the first time, an interpretation of thermodynamic entropy, in terms of mechanical and probabilistic concepts. Its influence is far reaching in physics. It is a high point in the development of the atomic world view of matter and the kinetic theory of dilute gases. Even today, the Boltzmann equations are routinely used for calculating transport coefficients and modeling approaches to equilibrium quantitatively.

11.2 The Boltzmann equation

As we know from earlier chapters, a statistical-mechanical system is described classically by the probability distribution, $\rho(\Gamma)$, where Γ is the vector of all canonical coordinates and momenta. Unfortunately, this function contains too much information and is too complicated to handle practically. This multi-variable function has only a formal value in the theory of nonequilibrium statistical mechanics. In order to reduce the complexity, we consider a reduced description, the probability (or actually the particle density) of one particle,

$$f(\mathbf{r}, \mathbf{p}, t) = N \int \rho(\mathbf{r}, \mathbf{p}, \mathbf{r}_2, \mathbf{p}_2, \dots, t) d^3\mathbf{r}_2 d^3\mathbf{p}_2 \cdots d^3\mathbf{r}_N d^3\mathbf{p}_N. \quad (11.1)$$

Here we consider a system of N-particles, and focus on the particle labelled 1 and integrate all the other degrees of freedom. The distribution f is normalized according to,

$$\int f(\mathbf{r}, \mathbf{p}, t) d^3 r d^3 \mathbf{p} = N. \tag{11.2}$$

Thus, $f d^3 \mathbf{r} d^3 \mathbf{p}$ gives the number of atoms having position $\mathbf{r}$ and momentum $\mathbf{p}$ in a small hyper-volume $d^3 r d^3 \mathbf{p}$ of the one particle phase space. This one-particle phase space is a 6-dimensional space. It is a smaller space than the $6N$-dimensional full phase space. Given the Hamiltonian, it is possible to write down an exact equation for f, based on the general Liouville equation. However, such an equation is not closed in the sense that it requires us to define other two particle or three particle analogues of the f function. The set of infinite series of equations is known as the BBGKY (Bogoliubov-Born-Green-Kirkwood-Yvon) hierarchy. If we use certain mean-field approximations, and try to close it at the lowest order of f, we can derive the Boltzmann equation [Harris (2004)]. The approximation can be rigorously justified in the dilute limit known as the Grad limit. Here, we will be content to give an intuitive and plausible argument for why the equation is valid.

First we note that if the particle in a force field $\mathbf{F}$ does not see the other particles (non-interacting) the Liouville equation is still valid, which is

$$\frac{Df}{Dt} = \frac{\partial f}{\partial t} + \frac{\mathbf{p}}{m} \cdot \frac{\partial f}{\partial \mathbf{r}} + \mathbf{F} \cdot \frac{\partial f}{\partial \mathbf{p}} = 0. \tag{11.3}$$

Here m is the mass of the particle, $\partial/\partial \mathbf{r}$ is the gradient operator with respect to space, and $\partial/\partial \mathbf{p}$ is with respect to momentum. The two gradient terms will be called the stream terms as they denote the change of position and momentum caused by the fact that the particle has a velocity and experience a force. Such motion preserves the phase space volume. The total derivative $Df/Dt = 0$ implies that if we follow the motion of the particles in a small phase space volume, the number of particles in this small hyper-volume is unchanged as they move.

However, the particles described by position $\mathbf{r}$ and momentum $\mathbf{p}$ collide with other particles. This causes scattering of the particles into and out of the phase space volume. The general Boltzmann equation takes the form

$$\frac{\partial f}{\partial t} + \frac{\mathbf{p}}{m} \cdot \frac{\partial f}{\partial \mathbf{r}} + \mathbf{F} \cdot \frac{\partial f}{\partial \mathbf{p}} = \left(\frac{\partial f}{\partial t} \right)_{\text{colli}}. \tag{11.4}$$

Here the right-hand side is called the collision rate. The form of this collision rate depends on the detailed mechanism of collision. In the simplest possible approximation to this term, we can give a relaxational term, known as the single-mode relaxation form,

$$\left(\frac{\partial f}{\partial t}\right)_{\text{colli}} = -\frac{f - f_0}{\tau}, \qquad (11.5)$$

where τ is the relaxation time which could depend on the momentum $\mathbf{p}$, f_0 is the equilibrium distribution, for example, the Maxwell distribution. The effect of the collision is to relax towards equilibrium at a time scale given by τ.

In the original formulation of Boltzmann, he considered binary collisions. Two particles of momenta $\mathbf{p}$ and $\mathbf{p}_1$ approach each other and scatter away with new momenta $\mathbf{p}'$ and $\mathbf{p}'_1$. We assume the collision is elastic and momentum and energy are conserved.

$$\mathbf{p} + \mathbf{p}_1 = \mathbf{p}' + \mathbf{p}'_1, \quad \epsilon(\mathbf{p}) + \epsilon(\mathbf{p}_1) = \epsilon(\mathbf{p}') + \epsilon(\mathbf{p}'_1), \qquad (11.6)$$

here $\epsilon(\mathbf{p}) = p^2/(2m)$ is the kinetic energy of the particle. Such a process can be described in mechanics by the scattering cross section. The differential cross-section $\sigma(\theta, \phi)$ is defined such that, if there is an incoming particle flux of I, the number of incident atoms scattered per second into the solid angle $d\Omega = \sin\theta d\theta d\phi$ in the outgoing direction (θ, ϕ) is given by $I\sigma(\theta, \phi)d\Omega$. For the binary collision, the collision term takes the following form:

$$\left(\frac{\partial f}{\partial t}\right)_{\text{colli}} = -\int d^3\mathbf{p}_1 d\Omega |\mathbf{v}_1 - \mathbf{v}|\sigma(\theta, \phi)(ff_1 - f'f'_1). \qquad (11.7)$$

Here $\mathbf{v} = \mathbf{p}/m$, $\mathbf{v}_1 = \mathbf{p}_1/m$ is the velocity vectors of the two particles, and we have introduced short-hand notation, $f = f(\mathbf{r}, \mathbf{p}, t)$, $f_1 = f(\mathbf{r}, \mathbf{p}_1, t)$, and similarly for the primed version. The first ff_1 term represents the scattering away from the state $\mathbf{p}$, causing a decrease of the particle numbers, while the second term is for the reverse process, causing an increase of the particle numbers in the state $\mathbf{p}$. In equilibrium or steady state, the two terms must balance each other, giving zero collision rate. The momenta before and after collisions, $\mathbf{p}$, $\mathbf{p}_1$, and $\mathbf{p}'$, $\mathbf{p}'_1$, have 12 degrees of freedom. The conservation laws take away 4 degrees of freedom. This still leaves with 8 degrees of freedom, which is $\mathbf{p}$, $\mathbf{p}_1$, and $d\Omega$. While $\mathbf{p}$ is fixed and given, the other remaining variables are integrated.

11.3   *H*-theorem

Boltzmann introduced a quantity defined as

$$H = \int f \ln f \, d^3\mathbf{r} d^3\mathbf{p}. \qquad (11.8)$$

Advanced Statistical Mechanics

This formula can be motivated from the counting of microscopic states, as discussed in Sec. 2.9. We can identify the entropy of a dilute gas as $S = -k_B H$. If f satisfies the Boltzmann equation, we can show that H is a non-increasing function of time, $dH/dt \leq 0$. This has precisely the property of entropy as given by the second law of thermodynamics.

In order to prove the H-theorem, we write the collision term in a more symmetric form as,

$$\left(\frac{\partial f}{\partial t}\right)_{\text{colli}} = -\int d^3\mathbf{p}_1 d^3\mathbf{p}' d^3\mathbf{p}'_1 \sigma(\mathbf{p}\mathbf{p}_1|\mathbf{p}'\mathbf{p}'_1)(ff_1 - f'f'_1). \quad (11.9)$$

Here we integrate all the momentum variables except $\mathbf{p}$. Since the scattering must conserve energy and momentum, the scattering cross-section function with four momentum arguments must contain four delta functions. The 6-dimensional integral for the primed variables reduces to the solid angle integration. We explore the symmetries of the scattering cross sections — they are exchange symmetric and time-reversal symmetric:

$$\mathbf{p} \leftrightarrow \mathbf{p}_1, \mathbf{p}' \leftrightarrow \mathbf{p}'_1, \quad \sigma(\mathbf{p}\mathbf{p}_1|\mathbf{p}'\mathbf{p}'_1) = \sigma(\mathbf{p}_1\mathbf{p}|\mathbf{p}'_1\mathbf{p}'), \quad (11.10)$$

$$\mathbf{p} \leftrightarrow \mathbf{p}', \mathbf{p}_1 \leftrightarrow \mathbf{p}'_1, \quad \sigma(\mathbf{p}\mathbf{p}_1|\mathbf{p}'\mathbf{p}'_1) = \sigma(\mathbf{p}'\mathbf{p}'_1|\mathbf{p}\mathbf{p}_1). \quad (11.11)$$

We now compute the rate of change,

$$\frac{dH}{dt} = \int \frac{Df \ln f}{Dt} d^3r d^3\mathbf{p} = \int (1 + \ln f) \frac{Df}{Dt} d^3r d^3\mathbf{p}. \quad (11.12)$$

We note that the integration of the total derivative over the whole phase space and of the partial derivative are the same as the extra stream terms can be written as a divergence which vanishes. Substituting the right-hand side of the Boltzmann equation into the above equation, we have

$$\frac{dH}{dt} = \int d^3r d^3\mathbf{p} d^3\mathbf{p}_1 d^3\mathbf{p}' d^3\mathbf{p}'_1 \sigma (1 + \ln f)(f'f'_1 - ff_1). \quad (11.13)$$

In the above, all the momenta are dummy integration variables. We make three kinds of permutation/exchange and then add to the original and divide by 4. The four alternative but equivalent expressions are: (1) the original term $\phi_1 = (1 + \ln f)(f'f'_1 - ff_1)$, (2) swapping the role of the incoming two particles, $\mathbf{p} \leftrightarrow \mathbf{p}_1$, $\phi_2 = (1 + \ln f_1)(f'_1 f' - f_1 f)$, (3) swapping prime and non-prime from the original expression, $\phi_3 = (1 + \ln f')(ff_1 - f'f'_1)$, (4) swapping prime and non-prime from case (2), $\phi_4 = (1 + \ln f'_1)(f_1 f - f'_1 f')$. We add them all and divide by 4. Since the differential cross section

σ is invariant under the permutation of particles and time reversal, it is a common factor, we have

$$\frac{dH}{dt} = \frac{1}{4} \int d^3\mathbf{r} d^3\mathbf{p} d^3\mathbf{p}_1 d^3\mathbf{p}' d^3\mathbf{p}_1' \sigma \, (B - A) \ln \frac{A}{B} \le 0. \qquad (11.14)$$

Here we have defined $A = f f_1$, $B = f' f_1'$. Since $\sigma \ge 0$, $A \ge 0$, and $B \ge 0$, the expression $(B - A) \ln(A/B) \le 0$ for any choice of A and B. The inequality is proved.

If we start from some arbitrary initial distribution f, as time develops, entropy will increase and eventually reach a maximum. Then what is the equilibrium distribution? We see for the Boltzmann equation, the collision conserves energy and momentum, the most general equilibrium solution is of the form

$$f(\mathbf{r}, \mathbf{p}, t) = \text{const}\, e^{-\beta\left(\epsilon(\mathbf{p}) + V(\mathbf{r}) - \mathbf{v}_0 \cdot \mathbf{p}\right)}. \qquad (11.15)$$

Here β is $1/(k_B T)$, $\mathbf{v}_0$ is an average drift velocity of the gas, and $V(\mathbf{r})$ is the single particle potential such that the force acting on the particle is given by $\mathbf{F} = -\partial V/\partial \mathbf{r}$. If we substitute this expression into the collision term, we find the collision rate is 0, due to energy and momentum conservation. The left-hand side is also 0 if we set $\mathbf{v}_0 = 0$ or $V = 0$ (why?).

11.4 Microscopic laws and irreversibility

In the last section of the book, we discuss a foundation question of macroscopic irreversibility and microscopic reversibility. We know all microscopic laws of dynamics are time-reversal symmetric. Newton's 2nd law together with an instantaneous force (such as the gravitational force) is time-reversal symmetric, in the sense that Newton's equation of motion is invariant under $t \to -t$. One cannot distinguish the motion of a projectile played forward or backward. The alternative formulation of Hamiltonian dynamics is also time reversal symmetric. Here, if we start from a phase space point $\Gamma_0 = (q_0, p_0)$, after an elapse of time t, the particle moves to $\Gamma = (q, p) = e^{t(\cdot, H)} \Gamma_0$. A reverse trajectory in physical space, still evolving forward in time is, $e^{t(\cdot, H)}(q, -p) = (q_0, -p_0)$, that is, we need to reverse the momentum variables. In quantum mechanics, if $\Psi(t)$ is a solution of the Schrödinger equation, the time-reversed one, $\Psi^*(-t)$ is also a solution of the Schrödinger equation. For example, $\Psi(t) = e^{i(\mathbf{k} \cdot \mathbf{r} - \omega t)}$ is a plane wave moving with momentum $\hbar \mathbf{k}$. If we make the transformation of $t \to -t$, and then take the complex conjugate, the new wave function $\Psi^*(-t) = e^{i(-\mathbf{k} \cdot \mathbf{r} - \omega t)}$ represents a wave moving backwards.

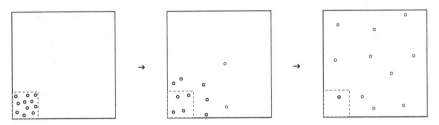

Fig. 11.1 Gas particles released from a small volume V_0 to a large volume V, increasing the entropy.

Consider a thought experiment of particles in a box, see Fig. 11.1. If we put one classical particle in a box, initially in a small corner of volume V_0, its motion is clearly reversible, as one cannot differentiate if the motion is played forward or backward. The same holds true if you have two or five particles. However, if the number of particles becomes very large, say, 10^{23} atoms, something qualitatively different happens. With any given reasonable observation time, we only observe the forward motion, never the backward motion, for the system as a whole. This is described by the entropy increase, according to Boltzmann's principle, $S = k_B \ln \Omega$. Here the entropies are specified by the macroscopic constraints, such as the initial and final volume V_0 and V, and not the microscopic detail, Γ. We recall the Sackur-Tetrode formula, then the entropy increase is given by $\Delta S = N k_B \ln(V/V_0)$.

Poincaré in 1892 presented a rigorous mathematical theorem which says that the configuration Γ will eventually returns as close as one care to ask for, to the initial state Γ_0. This is known as the Poincaré recurrence theorem. The theorem is based on the fact that the physically accessible phase space volume is finite and Liouville's theorem says the mapping of the phase space volume is preserved. However, this theorem does not have practical physical consequence as the recurrent time is usually very large, on the order of V^N, which diverges to infinity as both V and N become macroscopically large.

In this course, we also encounter dynamic equations that are manifestly time-reversal asymmetric, such as the diffusion equation, the Fokker-Planck equation, and the Boltzmann equation. For example, the Boltzmann equation predicts an entropy increase with time, consistent with the second law of thermodynamics. How can such equations be rigorously derived if the underlining microscopic picture is time-reversal symmetric? Such questions have been asked and answered in the literature [Lebowitz (2007)].

My understanding is that some limiting process must be taken place (such as $N \to \infty$), in order for this to happen.

Finally, it is tempting to generalize Boltzmann's entropy formula for the full system, known as Gibbs entropy,

$$S_G = -k_B \int \rho \ln \rho \, d\Gamma, \tag{11.16}$$

and consider it valid for nonequilibrium systems. Although this formula is certainly correct for equilibrium systems for the entropy, it fails to be an increasing function of time in nonequilibrium situation. Since the full probability distribution ρ satisfies the Liouville equation, we can easily show that $dS_G/dt = 0$. The entropy calculated with the Gibbs formula never increases in time. If the entropy has to do with information, then ρ carries the full information. Information is not lost in ρ, thus the entropy also does not change if defined according to ρ. This course started with the discussion of thermodynamic entropy following Callen, it is appropriate we end it here with that of Boltzmann and Gibbs.

11.5 Supplementary reading — electron transport, Boltzmann vs. Kubo-Greenwood

We discuss the Boltzmann equation description for electron transport [Ziman (1960)]. Here we consider a particularly simple treatment of electron conduction, through the single mode relaxation time approximation, and also show the connection of the Kubo-Greenwood formula to the Boltzmann equation result. With a single mode relaxation time approximation to the collision term, the Boltzmann equation takes a very simple form:

$$\frac{\partial f}{\partial t} + \frac{\mathbf{p}}{m} \cdot \frac{\partial f}{\partial \mathbf{r}} + \mathbf{F} \cdot \frac{\partial f}{\partial \mathbf{p}} = -\frac{f - f_0}{\tau}. \tag{11.17}$$

Here we assume the distribution function $f = f(\mathbf{r}, \mathbf{k}, t)$ is a function of time t, position $\mathbf{r}$, and wavevector $\mathbf{k} = \mathbf{p}/\hbar$. The relaxation time could be just a constant or it can depend on the wavevector $\mathbf{k}$ (or equivalently momentum $\mathbf{p}$). f_0 is the equilibrium distribution function, which is the Fermi function at chemical potential μ,

$$f_0 = \frac{1}{e^{\beta(\epsilon_\mathbf{k} - \mu)} + 1}. \tag{11.18}$$

$\epsilon_\mathbf{k}$ is the electron dispersion relation. For simple metals such as Al, it is well-described by free electron of $\hbar^2 k^2/(2m)$, but for 2D graphene, it is given by a massless Dirac dispersion of $\pm v_F \hbar |\mathbf{k}|$. The force $\mathbf{F}$ is the external force which we assume is due to the applied electric field, given by $\mathbf{F} = -e\mathbf{E}$. We assume $\mathbf{E}$ is small, and we look for the first order deviation from equilibrium. We also look for the steady state, therefore f will be time-independent and also homogeneous in a volume V, thus also position independent. The only variable in f will be $\mathbf{k}$. The Boltzmann equation is immediately 'solved' by multiplying the relaxation time throughout, given:

$$\delta f = f - f_0 \approx e\tau \mathbf{E} \cdot \mathbf{v_k} \frac{\partial f_0}{\partial \epsilon_\mathbf{k}}. \qquad (11.19)$$

Here in the partial derivative term, we have used the chain rule of differentiation and replaced f by the equilibrium value of f_0. The error introduced is of higher order. Here $\mathbf{v_k} = \partial \epsilon_\mathbf{k}/\partial(\hbar \mathbf{k})$ is the electron group velocity at the wavevector $\mathbf{k}$.

When the system is at equilibrium, positive velocity electrons and negative velocity electrons are equally likely, we don't have overall electronic current. The applied electric field causes the distribution to shift in the direction opposite to the field. This imbalance is reflected in the above formula. The electric current density is then calculated by

$$\mathbf{j} = (-2e) \int \frac{d^3\mathbf{k}}{(2\pi)^3} \mathbf{v_k} \delta f = 2 \int \frac{d^3\mathbf{k}}{(2\pi)^3} e^2 \tau \mathbf{v_k} \mathbf{v_k} \cdot \mathbf{E} \left(-\frac{\partial f_0}{\partial \epsilon_\mathbf{k}} \right). \qquad (11.20)$$

Here the factor of 2 takes into account the spin degeneracy of the electrons. We use the dyadic notation, thus $\mathbf{v_k}\mathbf{v_k}$ is a rank-2 tensor with elements $v_\mathbf{k}^\alpha v_\mathbf{k}^\beta$ for the $\alpha\beta$ component. The tensor (or matrix) is dotted (or scalar product) into the vector of field $\mathbf{E}$. We can read off the conductivity (the diagonal part in z direction) for an isotropic system as

$$\sigma = 2 \int \frac{d^3\mathbf{k}}{(2\pi)^3} e^2 \tau (v_\mathbf{k}^z)^2 \left(-\frac{\partial f_0}{\partial \epsilon_\mathbf{k}} \right). \qquad (11.21)$$

Next, we demonstrate that the Kubo-Greenwood result of conductivity agrees with the Boltzmann equation result above. To show this, we note that the trace is invariant with respect to a unitary transform. Instead of performing the trace in real space lattice sites j, we can perform the trace in the wavevector $\mathbf{k}$ space. For simplicity we assume the system has one site per unit cell, thus it will have only one band in the reciprocal space. In $\mathbf{k}$ space, the velocity matrix defined by Eq. (10.26) is diagonal and is given by the group velocity $\mathbf{v_k} = \partial \epsilon_\mathbf{k}/\partial(\hbar \mathbf{k})$. The retarded Green's function is also diagonal in **k** space, given by

$$G^r_{\mathbf{k}}(E) = \frac{1}{E + i\eta - \epsilon_{\mathbf{k}}}. \tag{11.22}$$

The small number $\eta > 0$ is an important feature here. It could also be **k** dependent. As we will see it is related to the electron relaxation time τ by $\eta = \hbar/(2\tau)$. We can also take into account the degeneracy of the spin, thus, trace over site j (and spin σ) becomes

$$\text{Tr}(\cdots) = 2 \sum_{\mathbf{k}} (\cdots) = 2V \int \frac{d^3\mathbf{k}}{(2\pi)^3} (\cdots). \tag{11.23}$$

We may think of **k** as discrete with values spaced in each direction by $2\pi/L$ where L is the linear dimension with volume $V = L^3$. Alternatively, we can also take the large volume limit, which converts the discrete sum into an integral of **k** over the first Brillouin zone. With this transformation, the Kubo-Greenwood formula becomes

$$\sigma = 2 \int \frac{d^3\mathbf{k}}{(2\pi)^3} \frac{\hbar e^2}{\pi} \left(-\frac{\partial f_0}{\partial E} \right)_{E=\epsilon_{\mathbf{k}}} (v^z_{\mathbf{k}})^2 \int dE \left(\text{Im} G^r_{\mathbf{k}}(E) \right)^2. \tag{11.24}$$

Using the explicit expression for the retarded Green's function, and the spectrum function $A_{\mathbf{k}}(E) = i(G^r_{\mathbf{k}}(E) - G^a_{\mathbf{k}}(E)) = -2\text{Im} G^r_{\mathbf{k}}(E)$, with

$$\text{Im} G^r_{\mathbf{k}}(E) = \frac{-\hbar/(2\tau)}{(E - \epsilon_{\mathbf{k}})^2 + \left(\frac{\hbar}{2\tau} \right)^2}, \tag{11.25}$$

we find,

$$\sigma = \frac{2\hbar e^2}{\pi} \int \frac{d^3\mathbf{k}}{(2\pi)^3} \left(-\frac{\partial f_0}{\partial \epsilon_{\mathbf{k}}} \right) (v^z_{\mathbf{k}})^2 \left(\frac{\hbar}{2\tau} \right)^2 \int_{-\infty}^{+\infty} \frac{dE}{\left[(E - \epsilon_{\mathbf{k}})^2 + \left(\frac{\hbar}{2\tau} \right)^2 \right]^2}.$$

Using the integral formula

$$\int_{-\infty}^{+\infty} \frac{dx}{(x^2 + 1)^2} = \frac{\pi}{2}, \tag{11.26}$$

with a change of integration variable to $x = 2\tau(E - \epsilon_{\mathbf{k}})/\hbar$, we obtain an expression identical to the Boltzmann result, Eq. (11.21). We also recover the Drude formula, $\sigma = e^2 n\tau/m$, if we assume a parabolic dispersion, $\epsilon_{\mathbf{k}} = \hbar^2 k^2/(2m)$, and take the low-temperature limit, $T \to 0$.

Problems

Problem 11.1. *Work out the relationship between the four-momentum cross section $\sigma(\mathbf{p}\mathbf{p}_1|\mathbf{p}'\mathbf{p}_1')$, and the two-parameter cross section with the solid angle, $\sigma(\theta, \phi)$.*

Problem 11.2. *Prove that the Gibbs entropy, $S_G = -k_B \int \rho \ln \rho \, d\Gamma$, satisfies, $dS_G/dt = 0$, if ρ satisfies the Liouville equation.*

Problem 11.3. *Consider a simplified Boltzmann equation of the form*

$$\frac{\partial f}{\partial t} + \frac{\mathbf{p}}{m} \cdot \frac{\partial f}{\partial \mathbf{r}} = -\frac{f - f_{\text{eq}}}{\tau},$$

where f_{eq} is the local equilibrium distribution having the property $\langle \ln f_{\text{eq}} \rangle_f = \langle \ln f_{\text{eq}} \rangle_{f_{\text{eq}}}$, m is mass of the particle, $\tau > 0$ is the relaxation time constant, and the distribution function $f = f(\mathbf{r}, \mathbf{p}, t)$ is a function of three-dimensional position $\mathbf{r}$, momentum $\mathbf{p}$, and time t. The Boltzmann H-function is defined as $H = \int d^3r d^3p \, f \ln f$. Prove the H-theorem: $dH/dt \leq 0$.

Problem 11.4. *Consider the Boltzmann equation for electrons in a metal driven under a high frequency electric field in x direction, $\mathbf{E} = \hat{x}Ee^{-i\omega t}$. Due to this sinusoidal external drive, the time-dependent solution of the Boltzmann equation in steady state also has the same frequency, i.e., $fe^{-i\omega t}$. We assume the field amplitude E is small so that the first order perturbation in E is valid. Using the constant relaxation time τ approximation for the collision term, show that the AC conductivity is given by*

$$\sigma(\omega) = e^2 \left(\frac{n}{m} \right) \frac{1}{1/\tau - i\omega},$$

where we have defined the ratio of the electron density to the effective mass as

$$\left(\frac{n}{m} \right) = 2 \int v_x^2 \left(-\frac{\partial f_0}{\partial \epsilon_{\mathbf{k}}} \right) \frac{d^3\mathbf{k}}{(2\pi)^3},$$

where the factor 2 is for spin degeneracy, v_x is the x-component of the group velocity, $\epsilon_{\mathbf{k}}$ is the dispersion relation of the electron, and $f_0 = 1/(e^{(\epsilon_{\mathbf{k}} - \mu)/(k_B T)} + 1)$ is the Fermi function at chemical potential μ and temperature T.

Problem 11.5. *The electron-phonon system can be described by a pair of Boltzmann equations for electron distribution f and phonon average* *occupation N in the form*

$$\frac{\partial f}{\partial t} + \mathbf{v}_k \cdot \frac{\partial f}{\partial \mathbf{r}} + \mathbf{F} \cdot \frac{\partial f}{\partial \hbar \mathbf{k}} = \left(\frac{\partial f}{\partial t}\right)_{\text{colli}},$$

$$\frac{\partial N}{\partial t} + \mathbf{v}_q \cdot \frac{\partial N}{\partial \mathbf{r}} = \left(\frac{\partial N}{\partial t}\right)_{\text{colli}},$$

here $\mathbf{v}_k$ is the electron group velocity for the mode $k = (n, \mathbf{k})$, n is the band index and $\mathbf{k}$ is the wavevector, similarly $\mathbf{v}_q$ is phonon group velocity at branch λ and wavevector $\mathbf{q}$, with the short-hand notation $q = (\lambda, \mathbf{q})$, as well as $f = f_k = f(t, \mathbf{r}, n, \mathbf{k})$ and $N = N_q = N(t, \mathbf{r}, \lambda, \mathbf{q})$. We ignore "force" on phonon due to lattice distortion. The collision term for the electron scattering due to phonons is given by

$$\left(\frac{\partial f_k}{\partial t}\right)_{\text{colli}} = \frac{1}{M} \sum_{k'} \left(S_{kk'}(1 - f_k)f_{k'} - S_{k'k}(1 - f_{k'})f_k\right),$$

$$S_{kk'} = \frac{2\pi}{\hbar} \sum_q |g_{kk'}^q|^2 \Delta(\mathbf{k} - \mathbf{k}' - \mathbf{q})\Big(N_q\,\delta(\epsilon_k - \epsilon_{k'} - \hbar\omega_q)$$

$$+ (N_{-q} + 1)\,\delta(\epsilon_k - \epsilon_{k'} + \hbar\omega_{-q})\Big).$$

Here $g_{kk'}^q$ is the electron-phonon scattering matrix element, ϵ_k is electron dispersion relation (band structure), and first Δ is the Kronecker delta modulo reciprocal lattice vector $\mathbf{G}$ for the quasi-momentum conservation, and second δ is the Dirac delta for energy conservation. M is the number of unit cells. The notation $-q$ means $(\lambda, -\mathbf{q})$. The collision term for phonon is

$$\left(\frac{\partial N_q}{\partial t}\right)_{\text{colli}} = \frac{2\pi}{\hbar M} \sum_{kk'} |g_{kk'}^q|^2 \Delta(\mathbf{k}' - \mathbf{k} + \mathbf{q})\delta(\epsilon_k - \epsilon_{k'} - \hbar\omega_q)$$

$$\times \Big[(1 - f_{k'})f_k(N_q + 1) - (1 - f_k)f_{k'}N_q\Big].$$

(a) Show that the Fermi and Bose equilibrium distributions

$$f_k^0 = \frac{1}{e^{\beta(\epsilon_k - \mu)} + 1}, \qquad N_q^0 = \frac{1}{e^{\beta\hbar\omega_q} - 1},$$

satisfy the equations in equilibrium (time-independent and no external force) and give 0 for the collision terms. (b) If we define the energy of the combined system as

$$E = \frac{1}{V} \int d^3\mathbf{r} \left(\sum_k \epsilon_k f_k + \sum_q \hbar\omega_q N_q\right),$$

show that energy is conserved, i.e., $dE/dt = 0$, when Boltzmann equations are satisfied. Here V is the volume of the system. Similarly, show that the *total number of electrons is conserved,* $dN_e/dt = 0$ *with*

$$N_e = \frac{1}{V} \int d^3\mathbf{r} \sum_k f_k.$$

(c) *Prove the H-theorem,* $dH/dt \leq 0$, *here we define*

$$H = \frac{1}{V} \int d^3\mathbf{r} \left\{ \sum_k \left(f \ln f + (1-f)\ln(1-f) \right) \right.$$

$$\left. + \sum_q \left(N \ln N - (N+1)\ln(N+1) \right) \right\}.$$

(d) *In the linear response regime when the deviation from equilibrium is small, we can write*

$$f = f^0 - \frac{\partial f^0}{\partial \epsilon}\Phi, \quad N = N^0 - \frac{\partial N^0}{\partial(\hbar\omega)}\Psi, \qquad (11.27)$$

Show that the collision terms can be simplified as

$$\left(\frac{\partial f_k}{\partial t} \right)_{\text{colli}} = \frac{\beta}{M} \sum_{k',q,\sigma=\pm 1} P^q_{kk'}(\sigma)(\Phi' - \Phi + \sigma\Psi),$$

$$\left(\frac{\partial N_q}{\partial t} \right)_{\text{colli}} = -\frac{\beta}{M} \sum_{kk'} P^q_{kk'}(+1)(\Phi' - \Phi + \Psi),$$

to first order in the small quantities Φ *and* Ψ. *Here we have used short-hand notations,* $\Phi = \Phi_k$, $\Phi' = \Phi_{k'}$, *and* $\Psi = \Psi_q$. *And the coefficient* $P^q_{kk'}(\sigma)$ *is defined by*

$$P^q_{kk'}(+1) = \frac{2\pi}{\hbar}|g^q_{kk'}|^2 \Delta(\mathbf{k} - \mathbf{k}' - \mathbf{q})\delta(\epsilon - \epsilon' - \hbar\omega)N^0(1-f^0)f^{0'},$$

$$P^q_{kk'}(-1) = \frac{2\pi}{\hbar}|g^q_{k'k}|^2 \Delta(\mathbf{k} - \mathbf{k}' + \mathbf{q})\delta(\epsilon - \epsilon' + \hbar\omega)(N^0+1)(1-f^0)f^{0'}.$$

Again note the obvious short-hand notation, $\epsilon = \epsilon_k$, $\epsilon' = \epsilon_{k'}$, $\omega = \omega_q$, *and* $f^0 = f^0(\epsilon_k)$, $(f^0)' = f^0(\epsilon_{k'})$, $N^0 = N^0(\omega_q)$. (e) *The single mode relaxation time approximation is obtained by ignoring the off-diagonal terms in the collision terms. The relaxation times are defined by* $\left(\frac{\partial f}{\partial t} \right)_{\text{colli}} = -(f - f^0)/\tau$ *for electron, and* $\left(\frac{\partial N}{\partial t} \right)_{\text{colli}} = -(N - N^0)/\tau_p$ *for phonon. From these definitions determine the explicit expressions for the relaxation times.*

Answers to Selected Problems

1.1 (a) The state changes slowly in such a way that at each moment, the state is in thermal equilibrium. (b) The point of temperature and pressure where liquid water, ice, and vapor are in phase coexistence. (c) The change of energy with particle number under constant entropy and volume, $(\partial U/\partial N)_{S,V}$. (d) This makes temperature positive but it is not absolutely necessary. (e) Thermodynamic stability requires this. (f) In adiabatic processes entropy cannot decrease. (g) 0-th law of thermodynamics — systems that are in equilibrium with each other have the same temperature. (h) Callen's four postulates. (i) an adiabatic wall cannot pass heat, diathermal wall can pass heat but not particles, semi-permeable can allow certain type of particles passing through. A closed system is described by a Hamiltonian, and open system cannot, and isolated system means both energy and particle number cannot change.

1.2 We start from $TdS = dU + pdV$. This is $dS = \frac{C_V}{T}dT + Nk_B dV/V$. Take an arbitrary path in (T, V) space, we obtain $S = \frac{3}{2}Nk_B \ln T + Nk_B \ln V + \text{const}$. The constant part contains N dependence. One gets this by the requirement that S must be a homogeneous function of degree one in N. Use $U = 3Nk_B T/2$ to replace T, we obtain

$$S = Nk_B \ln \left(\frac{VU^{3/2}}{N^{5/2}}\right) + S_0(N).$$

Unfortunately, thermodynamics cannot fix the last constant term.

1.3 Using the result in problem 1.2, adiabatic process is given by $VU^{3/2} = \text{const}$, or $V_2 U_2^{3/2} = V_1 U_1^{3/2}$. So the internal energy change is $U_2/U_1 = (V_1/V_2)^{2/3} = 1/2^{2/3}$, or $U_2 - U_1 = (1/2^{2/3} - 1)U_1 \approx -0.37U_1$. Since T is proportional to U for ideal gas, we also have $T_2 - T_1 \approx -0.37\,T_1$.

In an adiabatic process $Q = 0$, so by the first law of thermodynamics, $W = U_2 - U_1 = -0.37\,U_1$. Negative sign means the system is doing work to outside.

1.4 Using the Maxwell equation and the fact that temperature T is a function of the energy density $u = U/V$ only, we find $\partial(u/(3T))/\partial U|_V = \partial(1/T)/\partial V|_U$. View $T = T(u)$, and $u = U/V$, we evaluate the partial derivatives and obtain a differential equation $4uT'(u)/T = 1$. The solution is $u = aT^4$. Applying the Euler relation with chemical potential $\mu = 0$, we find $S = (U + pV)/T = \frac{4}{3}aT^3V = \frac{4}{3}(aV)^{1/4}U^{3/4}$.

1.5 (c) The four equations are obtained by the requirements that the entropy and temperature must be a continuous function at U_1 and U_2. (d) The latent heat is simply the energy jump $U_2 - U_1$. (e) For energy $U_1 < U < U_2$, it can be interpreted as mixture of pure states at energy U_1 and U_2 at the same temperature $T = 1/b$.

1.6 (a) This is equivalent to the existence of a functional form $F(x, y, z) = 0$, so that x is an implicit function of y and z, etc. From this we get the differential $dF = \partial F/\partial x\, dx + \partial F/\partial y\, dy + \partial F/\partial z\, dz = F_x dx + F_y dy + F_z dz = 0$. Then the partial derivatives fixing one of the variables mean that variable has 0 differential, so we obtain $(\partial x/\partial y)_z = -F_y/F_x$, $(\partial y/\partial z)_x = -F_z/F_y$, $(\partial z/\partial x)_y = -F_x/F_z$. Multiplying the three factors, we obtain -1. (b) Let $x = V$, $y = T$, and $z = p$, using the result in (a), we obtain the identity with $B = -V(\partial p/\partial V)_T$. Here we have used $\partial V/\partial p = 1/(\partial p/\partial V)$ fixing T.

1.7 The heat capacity in general is $\delta Q/dT$, but $\delta Q = TdS$, thus $C_p = T\partial S/\partial T|_p = C_V + \partial S/\partial V|_T \partial V/\partial T|_p$. Here S is viewed as a function of (T, p) or (T, V), and $V = V(T, p)$. Using Maxwell's equation due to $dF = -SdT - pdV$, we find $\partial S/\partial V|_T = \partial p/\partial T|_V$. We replace it again by the triple product formula, $\partial p/\partial T|_V\, \partial T/\partial V|_p\, \partial V/\partial p|_T = -1$, and using the definitions for the thermal expansion coefficient and isothermal compressibility, we obtained the desired equation.

2.1 The micro-canonical number of states Ω and its consequence is given in the text. The canonical partition function of an ideal gas is $Z_N = (V/\lambda^3)^N/N!$, where $\lambda = h/\sqrt{2\pi mk_BT}$, and grand canonical partition is $\Xi = \sum_{N=0}^{\infty} Z_N e^{\beta\mu N} = \exp(e^{\beta\mu}V/\lambda^3)$. Here we have used the Taylor series of the exponential function, $e^x = \sum_{n=0}^{\infty} x^n/n!$. No matter which version of the statistical ensembles is used, the thermodynamic results are the same for the three cases. Details are omitted here.

2.2 (a) Yes, otherwise heat will flow from hot to cold and system is not in equilibrium. (b) Canonical is best as total particle number N, temperature and volume are fixed. (c) We focus on a single particle and calculate the probability p_1 that it is in V_1 (or $1 - p_1$ in V_2), this is just proportional to the Boltzmann factor, $e^{-\beta u(r)}$. So $N_1 = N/(1 + e^{-\beta u_0})$, and $N_2 = Ne^{-\beta u_0}/(1 + e^{-\beta u_0})$. (d) Applying the ideal gas law for each region, we get $p_1 = N_1 k_B T/V$ and $p_2 = N_2 k_B T/V$. Alternatively, one can also find the partition function first, $Z = z^N/N!$. $z = (2\pi m k_B T/h^2)^{3/2}(V_1 + V_2 e^{-\beta u_0})$. p_1 or p_2 is obtained by taking partial derivative with respective to V_1 or V_2 of the free energy $F = -k_B T \ln Z$, then set $V_1 = V_2 = V$. $p_1 \neq p_2$ and the extra force is due to the potential going from region 1 to 2.

2.3 (a) Using canonical distribution, $C_1 = dU/dT = d\langle H\rangle/dT = \frac{d}{dT}\int d\Gamma H e^{-\beta H}/Z$. The temperature dependence appears in two places, once in the exponential, and once in the partition function, so we get $C_1 = (\langle H^2\rangle - \langle H\rangle^2)/(k_B T^2)$. The fluctuation of energy can also be written as $\langle (H - \langle H\rangle)^2\rangle$, which is clearly non-negative so $C_1 \geq 0$. (b) In microcanonical ensemble, consider $1/T = dS/dU$. We differentiate with respect to U on both sides, and using the definition $C_2 = dU/dT$, we get $C_2 = -(T^2 d^2 S/dU^2)^{-1}$. (d) $C_1 = C_2$ in the thermodynamic limit, but for finite system, the concavity of entropy is not guaranteed. In particular, at a first-order phase transition point, $C_2 < 0$ is possible in a microcanonical ensemble.

2.6 (a) With one particle of dimension a moving in interval L, the available space is $Q_1 = L - a$ as minimum distance from the center of the molecule to the two walls on both sides is $a/2$. (b) and (c) in general we need to perform the integral

$$Q_N = \int_{a/2}^{L-(N-1+1/2)a} dx_N \int_{x_N+a}^{L-(N-2+1/2)a} dx_{N-1} \cdots$$
$$\times \int_{x_3+a}^{L-3a/2} dx_2 \int_{x_2+a}^{L-a/2} dx_1.$$

The integral can be performed recursively starting from x_1. One find $Q_N = (L - Na)^N/N!$, and the force is then $f = -\partial F/\partial L = k_B T N/(L - Na)$.

2.8 It is convenient that we start with the Gibbs volume entropy, $S = k_B \ln \Gamma$, here Γ is the phase space volume for $H \leq U$. The inverse temperature is obtained by taking a derivative with respect to the energy and then approximating the derivative by finite difference, we get

$$\frac{1}{T} = \frac{k_B}{\Gamma} \frac{\partial \Gamma}{\partial U} \approx \frac{k_B}{\Gamma} \frac{\Gamma(U + \Delta U) - \Gamma(U)}{\Delta U}.$$

Γ is also a function of the volume V and particle number N, we have omitted the arguments as they don't change. The difference in Γ is the volume of a shell bounded by $U < H < U + \Delta U$, which can be expressed by a surface integral,

$$\frac{\Delta \Gamma}{\Delta U} = \frac{1}{\Delta U} \int_{U < H < U + \Delta U} d\Gamma = \int_{H = U} d\sigma \frac{1}{|\nabla H|}.$$

Here we used the fact that energy change can be written as $\Delta U = \nabla H \cdot \Delta x = |\nabla H| \Delta x$, and $d\Gamma = d\sigma \Delta x$. Δx is a distance in phase space normal to the energy surface $H = U$, and $d\sigma$ is $(2N - 1)$ dimensional surface element. Now if we choose a vector u in the phase space such that $\nabla \cdot u = 1$, we can express the volume in phase space by a surface integral, $\int_{H < U} d\Gamma = \int \nabla \cdot u \, d\Gamma = \int_{H = U} u \cdot n \, d\sigma$, here $n = \nabla H / |\nabla H|$ is the unit norm of the surface, and ∇H is the gradient of the Hamiltonian with respect to the set of dynamic variables (p, q). Then

$$k_B T = \frac{\int u \cdot \nabla H \frac{d\sigma}{|\nabla H|}}{\int \frac{d\sigma}{|\nabla H|}} = \langle u \cdot \nabla H \rangle_{\text{microcaonical}}.$$

The surface integral weighted by $1/|\nabla H|$ is precisely the microcanonical ensemble average in the limit $\Delta U \to 0$.

2.9 (a) The integration over the phase space Γ can be changed as integration over energy with the density of states $N(E)$ per unit energy interval. By the Boltzmann principle, we can write $\exp(S(E)/k_B) = N(E)$. Then the values of the two-dimensional integrals over the volume V and E is dominated by the maxima (multiplied by a gaussian), so $-1/\beta \ln Z_G \approx -1/\beta \ln e^{S/k_B - \beta(E + pV)} = -TS + E + pV = G$. (b) The $d\Gamma$ integral gives $\int d\Gamma e^{-\beta H} = V^N (2\pi m/\beta)^{3N/2}$. Using the given formula for $x!$ we cancel the $N!$ in the denominator. Find $Z_G = (2\pi m/(\beta h^2))^{3N/2}(1/(\beta p))^N$. (c) Since $G(T, p, N) = -k_B T \ln Z_G$, and take derivative with respect to N we get $\mu = \partial G/\partial N = -1/\beta \ln[(2\pi m/(\beta h^2))^{3/2} 1/(\beta p)]$. Take derivative with respect to pressure p, we get $V = \partial G/\partial p = N/(\beta p)$. Or the ideal gas law $pV = Nk_B T$. (d) Entropy is obtained by taking partial derivative with respect to temperature T. We got two terms which can be simplified to the standard Sackur-Tetrode form: $S = Nk_B \ln[V/(N\lambda^3)] + 5/2 Nk_B$, where $\lambda = h/\sqrt{2\pi m k_B T}$ is the thermal wavelength.

3.7 (a) Since the system is quadratic, we can always diagonalize the system using normal mode coordinates. Each quadratic form gives $(1/2)k_B T$ for energy, the total energy is $U = 2N(1/2)k_B T = Nk_B T$, the heat capacity is thus $C = dU/dT = Nk_B$. (b) Because $x_{N+1} = 0$, we must have $\sin(q(N+1)) = 0$, or $(N+1)q = k\pi$, or $q = k\pi/(N+1)$, $k = 1, 2, \ldots, N$. The value k specifies the N linearly independent solutions of normal mode vibrations. Substituting the solution $x_j = A\sin(qj)\cos(\omega_q t)$ into the equation of motion $m\ddot{x}_j = k(x_{j-1} - 2x_j + x_{j+1})$, we find $\omega_q^2 = (2k/m)(1 - \cos(q))$. (c) The energy of a set of quantum harmonic oscillators is $U = \sum_q (N_q + 1/2)\hbar\omega_q$, $N_q = 1/(e^{\beta\hbar\omega_q} - 1)$. The heat capacity is obtained by taking derivative with respect to temperature T, $C = dU/dT = \sum_q \hbar\omega_q dN_q/dT = k_B \sum_q (\beta\hbar\omega_q)^2 e^{\beta\hbar\omega_q}/(e^{\beta\hbar\omega_q} - 1)^2$, where q takes the value $k\pi/(N+1)$ for $k = 1, 2, \ldots, N$.

3.8 Define $\alpha = (\pi^2\hbar^2)/(2m)$, then $E_n = \alpha n^2/L^2$. The partition function is $Z = \sum_{n=1}^{\infty} e^{-\beta\alpha n^2/L^2} = x + x^4 + x^9 + \cdots$, where $x = e^{-\alpha\beta/L^2}$. For low temperatures, $\beta = 1/(k_B T)$ is large, and x is small, we just keep the first two terms in the Taylor expansion of x as a good approximation. For higher temperature, the summand varies slowly with n, we approximate the sum by integral, $Z \approx \int_0^{\infty} e^{-\alpha\beta n^2/L^2} dn = L\sqrt{2\pi m k_B T}/h$, (exactly the same result as for classical particle, i.e., $\int \int dx\, dp/h e^{-\beta p^2/(2m)}$) and the free energy is $F = -k_B T \ln Z$. Using force $f = -\partial F/\partial L$, we obtain, for (a) high temperature $f = k_B T/L$ (ideal gas law), (b) for low temperature $f = 2\alpha(1 + 3x^3 + \cdots)/L^3$. This is Casimir-force like.

3.9 (a) The spacing in wave vector k is $2\pi/L$, so the density in k-space is $L\,dk/(2\pi)$. Map energy ϵ to k, $k = \sqrt{2m\epsilon}/\hbar^2$, multiply by 4 (2 for spin, another 2 for $+k$ and $-k$ for each ϵ), we get $D(\epsilon) = (L/(\pi\hbar))\sqrt{2m/\epsilon}$. (b) $\ln \Xi = \int_0^{\infty} d\epsilon D(\epsilon) \ln[1 + e^{-\beta(\epsilon-\mu)}]$. (c) Let ϵ_F be the Fermi energy (μ at $T = 0$), then because variation occurs only near Fermi energy:

$$C = \frac{dU}{dT} = \int_0^{+\infty} d\epsilon(\epsilon - \epsilon_F) D(\epsilon) \frac{\partial f}{\partial T}$$

$$\approx D(\epsilon_F) \int_0^{+\infty} d\epsilon(\epsilon - \epsilon_F) \frac{\partial f}{\partial T}$$

$$\approx D(\epsilon_F) k_B^2 T \int_{-\infty}^{+\infty} \frac{x^2 e^x}{(e^x + 1)^2} dx = D(\epsilon_F) k_B^2 T \frac{\pi^2}{3}.$$

4.3 (a)
$$f(V) = \ln(V - Nb) + \frac{2aN}{k_B TV} - \frac{V}{V - Nb}.$$

(b) Expand $f(V)$ to 3rd order in small x, and equate $f(V_g) = f(V_l)$, we get $2f' + (1/3)f'''x^2 = 0$. After some simplification, we obtain the required result.

4.4 (c), (d) The heat capacity at constant pressure is

$$C_p = \frac{3}{2}Nk_B + \frac{Nk_B}{1 - \frac{2aN(V-Nb)^2}{k_B TV^3}} \approx \frac{3}{2}Nk_B + \frac{1}{3^{1/3}}Nk_B\left(\frac{T - T_c}{T_c}\right)^{-2/3}.$$

Note that we cannot assume V is a constant when pressure p is fixed.

4.5 (a) The temperature part is the momentum integrals, configuration partition function is the position integrals, so if $Z_N = P_N Q_N$, $P_N = (2\pi m/(\beta h^2))^{3N/2}/N!$. (b) $\Xi = \sum_{N=0}^{\infty} e^{\beta\mu N} Z_N = Q_0 + \alpha Q_1 + \alpha^2 Q_2/2 + \cdots$, where $\alpha = e^{\beta\mu}(2\pi m/(\beta h^2))^{3/2}$. $Q_0 = 1$, $Q_1 = V$, $Q_2 = V[e^{\beta\epsilon}4\pi/3(b^3 - a^3) + V - 4\pi b^3/3]$. (c) In grand canonical ensemble, $pV/(k_B T) = \ln\Xi$, and $N = 1/\beta\partial\ln\Xi/\partial\mu$. Equation of state is obtained after eliminating α from these two equations.

5.4 (a) Let the site of interest i be called 0 instead of i. Then $h_0 = h + J\sum_{j\in nn}\sigma_j$ here the sum is over the nearest neighbors of site 0. The cavity Hamiltonian is the remaining terms which is an Ising model with site 0 and the interaction with it removed, i.e., $H_{\text{cavity}} = -J\sum_{\langle ij\rangle, i,j\neq 0}\sigma_i\sigma_j - h\sum_{i\neq 0}\sigma_i$. (b) We have

$$\langle\sigma_0\rangle = \frac{1}{Z}\sum_{\sigma}\sigma_0\exp\left[-\beta(H_{\text{cavity}} - \sigma_0 h_0)\right].$$

We split the sum over all spins into sum over only σ_0 and sum over the rest of the spins, then

$$\langle\sigma_0\rangle = \frac{1}{Z}\sum_{\sigma_i, i\neq 0}\left(e^{\beta h_0} - e^{-\beta h_0}\right)e^{-\beta H_{\text{cavity}}}.$$

Now we multiply the summand by $1 = \sum_{\sigma_0} e^{\beta\sigma_0 h_0}/(e^{\beta h_0} + e^{-\beta h_0})$. The required identity is proved. (c) Expand the tanh function as a power series, then expand h_0 and the final expression as a power series in σ, using the assumption given, we can move the average sign inside, to get

$$\langle \sigma_i \rangle = \left\langle \tanh \left[\beta \left(h + J \sum_{j \in \text{nn of } i} \sigma_j \right) \right] \right\rangle$$

$$\approx \tanh \left[\beta \left(h + J \sum_{j \in \text{nn of } i} \langle \sigma_j \rangle \right) \right].$$

5.7 $h = \partial G / \partial M = 0$. This gives $M = \left(c + \sqrt{c^2 - 8ad(T - T^*)} \right)/(2d)$. T_c is determined by $G(M) = G(0)$. The magnetization drops to 0 at $M_0 = 2c/(3d)$ when $T_c = T^* + c^2/(9da)$.

6.1 (a) $P_{\sigma\sigma'} = \exp(\beta J \delta_{\sigma\sigma'})$, (b) $\lambda = e^{\beta J} + q - 1$, (c) $F = -k_B T N \ln \lambda$, $\xi^{-1} = \ln \left[(e^{\beta J} + q - 1)/(e^{\beta J} - 1) \right]$.

6.2 (a) Sum over the two spins at the outer vertices of the square, we can write the transfer matrix as PQ, here

$$P = \begin{pmatrix} z^4 + 1/z^4 + 2 & 4 \\ 4 & z^4 + 1/z^4 + 2 \end{pmatrix} \quad \text{and} \quad Q = \begin{pmatrix} z & 1/z \\ 1/z & z \end{pmatrix},$$

where $z = e^K = e^{\beta J}$. (b) There are two types of loops, loop over the squares with various combinatorial choices; the zigzag paths going from 1 to N and back to 1 again. This gives $Z = 2^{4N} \cosh^{5N}(K) \left[(1 + x^4)^N + (2x^3)^N \right]$. where $x = \tanh(K)$.

6.4 (a) Sum over the out spins 1, 2, 3 first, we find $(K = \beta J = J/(k_B T))$

$$Z_1 = \sum_{\sigma_0} \sum_{\sigma_1} e^{K\sigma_0\sigma_1} \sum_{\sigma_2} e^{K\sigma_0\sigma_2} \sum_{\sigma_3} e^{K\sigma_0\sigma_3}$$

$$= \sum_{\sigma_0} \left(e^{K\sigma_0} + e^{-K\sigma_0} \right)^3 = 2 \left(e^K + e^{-K} \right)^3.$$

For Z_2 we also sum over the outer spins first, we find $Z_2 = 2(e^K + e^{-K})^9$. (b) For the general case, we use high-temperature expansion. Since the tree graphs cannot have loops, all the $\tanh(x)$ parts are 0, and we only have the first term. $Z_N = 2^S \cosh^L(K)$ where $S = L + 1$ is the number of sites, and L is number of links, $L = 3(2^N - 1)$. (c) No phase transition as the partition function is the same, up to a constant factor, as the one-dimensional Ising model.

6.5 (b) $Z = 2^N (\cosh K)^L (1 + 2x^3 + x^4 + 2x^5 + 2x^6)$, $N = F^* = 6$, $L = L^* = 8$, $F = N^* = 4$.

6.6 (a) This is trickier than one thinks. The most obvious expansion variable is to use $e^{K\delta_{\sigma\sigma'}} = 1 + x\delta_{\sigma\sigma'}$, with $x = e^K - 1$. But this expansion doesn't have the Ising-like property in the sense averaging over spins produce 0, and the duality relation between high and low temperature expansion of the partition function cannot be established. Then when we expand, we get no delta factor (no bonds), one delta factor (one bond), and two delta factors (two bonds). Note that for the two delta factor case, the bond can be disconnected or connected, this gives two possible terms, thus the partition function is $Z = q^N + 2Nxq^{N-1} + N(2N-1)x^2q^{N-2} + O(x^3)$. Alternative to the delta expansion above is to use $q\delta_{\sigma\sigma'} - 1$. This has the property that a single term average to zero, so to get a nonzero value we must form a loop (as in the Ising case). We can write $e^{K\delta_{\sigma\sigma'}} = (e^K + q - 1)/q \left[1 + y(q\delta_{\sigma\sigma'} - 1)\right]$, here $y = (e^K - 1)/(e^K + q - 1) = x/(x+q)$ replacing x. With the variable y we can form loops of 4 sided and 6 sided, the first three terms are $Z = q^N \left[(e^K + q - 1)/q\right]^{2N}(1 + N(q-1)y^4 + 2N(q-1)y^6 + O(y^8))$. [But the trouble seems to be why the coefficient has the $q-1$ factor and not other expressions? Note the curious mathematical fact: if we define a $q \times q$ matrix P with elements $q\delta_{\sigma\sigma'} - 1$, we have $\text{Tr}(P^k) = q^k(q-1)$.]
(b) The low temperature expansion is straightforward, the ground state is q-fold degenerate with energy $-J2N$, and first excited states change one site, with energy higher by $4J$ and it is $Nq(q-1)$ fold degenerate, and next higher states flip two spin side-by-side, having energy higher than ground state by $6J$. So partition function is $Z = qe^{2NK} + q(q-1)Ne^{(2N-4)K} + 2Nq(q-1)e^{(2N-6)K} + \cdots$. We see the coefficients in part (a) and (b) are the same. This means duality is also true for Potts model and we can identify T_c by the equation $y = (e^K - 1)/(e^K + q - 1) = e^{-K}$. Solving the equation for K, we find $K_c = J/(k_B T_c) = \ln(1 + \sqrt{q})$.

7.1 (a) We need the basic definitions of critical exponents, $M \sim (-t)^\beta$, $\chi \sim t^{-\gamma}$, $h \sim M^\delta$. To get β, we set $h = 0$, and solve for M in terms of t, we find $M = (-t/b)^{1/2}$, so $\beta = 1/2$. Take derivative with respect to h for fixed t on both sides of the equation of states, consider the high temperature side so $M = 0$, we found $1 = a\chi t^\theta$, so $\gamma = \theta$. Lastly, to find δ, we set $t = 0$ in the equation of states, we identify $\delta = 1 + 2\theta$. Part (b) is easily verified. Note that this problem has nothing to do with Landau theory or mean-field free energy. The important point is that β and γ are defined at $h = 0$, and δ is defined at $t = 0$.

7.2 (a) Since $m = -\partial f/\partial h$, we can set $t = 0$, $b = L$, to get $f(0, h, L) = L^{-D} f(0, L^X h, 1)$. Differentiating with respect to h, then set $h = 0$, assuming the limit $h \to 0$ exists, we get $\Delta_1 = X - D$. (b) Differentiate one more time with respect to h of $f(0, h, L)$, we get $\Delta_2 = 2X - D$. (c) Differentiate 4 times, we get $\langle M^4 \rangle/N = L^{4X-D}$, and $\chi = \partial m/\partial h = \langle M^2 \rangle/(k_B T N)$ (assuming $t > 0$), where $N = L^D$, $M = \sum_i \sigma_i$. Taking the ratio of 4-th moment to the second moment squared, we find $\Delta_3 = 0$.

7.3 (a) $F(T, h) = G - hM$ where $h = \partial G/\partial M = (T - T_c) 2aM + 4bM^3$. Formally, F will be a function of T and h only, if we solve the above equation for M in terms of h. (b) at $T = T_c$, $h = 4bM^3$, so $F = bM^4 - hM \propto h^{4/3}$. (c) at $h = 0$, we have $(T - T_c)a + 2bM^2 = 0$, so $M \propto (T - T_c)^{1/2}$. $F = G = (T - T_c)aM^2 + bM^4 \propto (T - T_c)^2$. (d) set $t = 0$, $b^X h = 1$, we have $F(0, h) = h^{4/X} F(0, 1)$, compare with (b) $4/3 = 4/X$, so $X = 3$. Similarly, $F(t, 0) = t^{4/Y} F(1, 0)$, compare with (c) $4/Y = 2$, so $Y = 2$.

8.2 Let the cumulative distribution be $F(x) = P(\xi_1 + \xi_2 \le x) = \int_{\xi_1 + \xi_2 \le x} d\xi_1 d\xi_2 \cdot 1$. We obtain $F(x) = x^2/2$ if $0 \le x \le 1$, and $F(x) = 1 - (2 - x)^2/2$ if $1 \le x \le 2$. Differentiating, we find the probability density is $p(x) = F'(x) = x$ if $x \le 1$ and $2 - x$ if $x \ge 1$. It is a triangle-shaped distribution.

8.4 (a) We first work out the off-diagonal elements and the diagonal one is determined by the fact that the sum of a row normalizing to 1. Each element is $1/3$ times the Metropolis rate $\min(1, e^{-\beta \Delta E})$ thus we just need to work out the energy changes between states that differ by one spin. Let $x = e^{-\beta 6J}$, $y = e^{-\beta 2J}$, naming the states according to a binary coding, i.e., $(+ + +) \to 0$, $(+ + -) \to 1$, ..., $(- - -) \to 7$, we have, for the rate of $x/3$ transitions are $0 \to 1$, $0 \to 2$, $0 \to 4$. There are 9 terms with transition rate of $1/3$: $1 \to 0$, $2 \to 0$, $4 \to 0$, $3 \to 1$, $3 \to 2$, $5 \to 1$, $5 \to 4$, $6 \to 2$, $6 \to 4$. The remain 9 terms have transition rate of $y/3$: $1 \to 3$, $1 \to 5$, $2 \to 6$, $2 \to 3$, $4 \to 6$, $4 \to 5$, $0 \to 3$, $0 \to 5$, $0 \to 6$. All other entries other than the diagonals are 0. (b) It is just the Boltzmann distribution, $e^{-\beta H(\sigma)}/Z$, with $Z = 4 + 3e^{-2K} + e^{6K}$, $K = J/(k_B T)$.

8.5 Using L (with $LL^T = A^{-1}$), we do transform, by $x = Lz$, here z is independent gaussian which can be generated by the Box-Muller method. Then x will be multi-variable Gaussian distributed.

9.1 Use the definition of Bessel function,

$$J_n(x) = \frac{1}{2\pi} \int_{-\pi}^{+\pi} e^{i(n\tau - x\sin\tau)} d\tau.$$

9.3 (a) If $R = 0$, the solution is $v = v_0 e^{-\gamma t}$. Let $v(t) = A(t)e^{-\gamma t}$, substituting into the equation we can solve for A. Given solution, in the long time limit, $v(t) = \int_{-\infty}^{t} e^{\gamma(\tau - t)} R(\tau)/m \, d\tau$. (b) The velocity-random force correlation is

$$\langle v(t)R(t')\rangle = \left\langle \int_{-\infty}^{t} e^{\gamma(\tau - t)} \frac{R(\tau)}{m} R(t')d\tau \right\rangle$$

$$= \frac{C}{m} \int_{-\infty}^{t} \delta(\tau - t') e^{\gamma(\tau - t)} d\tau$$

$$= \begin{cases} 0 & \text{if } t < t', \\ \frac{C}{m} e^{-\gamma(t - t')} & \text{if } t > t'. \end{cases}$$

The fact that correlation is 0 when $t < t'$ reflects causality as future forces cannot influence the past.

9.4 (a) Integrate, we get $x(t) = (1/k) \int_0^t R(t')dt'$. Perform a two-dimensional integral over a square of $[0, t]^2$ and use the delta function correlation, we obtain, $\langle x(t)^2 \rangle = \frac{1}{k^2} \int_0^t dt_1 \int_0^t dt_2 \langle R(t_1)R(t_2)\rangle = Ct/k^2$. (b) Probability conservation means $\partial P/\partial t + \partial/\partial x(\dot{x}P) = 0$. Using the expression for the rate of x, and moving the second term to the right and solving the equation formally, then taken average over noise, or alternatively, using analogy to the standard Fokker-Planck equation, we obtain (skipping steps),

$$\frac{\partial P(x,t)}{\partial t} = \frac{C}{2k^2} \frac{\partial^2 P(x,t)}{\partial x^2},$$

which is a diffusion equation.

9.5 (a) $\mu = 1/(m\gamma)$. (b) $\partial P/\partial t = -\partial/\partial v \left[(-\gamma v + f/m)P \right] - \partial(vP)/\partial x + (\gamma k_B T)/m \partial^2 P/\partial v^2$. $P = P(t, x, v)$ is a function of three variables, t, x, v. (c) Set the left-hand side to 0, we can solve the (ordinary) different equation, given $P(x, v) \propto e^{-\beta(1/2mv^2 - xf)}$. Proportionality constant is fixed by normalization.

9.6 (a) The solution $x(t)$ is obtained by the method of variation of a constant, where we first let $R(t) = 0$, then $x(t) = Ae^{-ct}$ where $c = k/(m\gamma)$. Then we let $A \to A(t)$ and substitute back into the equation to obtain equation for $A(t)$. After integration we get

$$x(t) = A_0 e^{-ct} + \int_0^t \frac{R(s)}{m\gamma} e^{-c(t-s)} ds.$$

(b) We can follow the standard derivation of Zwanzig, but the equation is identical to the standard one if we identify x as velocity v, and some change of variables. So the Fokker-Planck equation is the same (skip the derivation)

$$\frac{\partial P}{\partial t} = \frac{k}{m\gamma} \frac{\partial(xP)}{\partial x} + \frac{k_B T}{m\gamma} \frac{\partial^2 P}{\partial x^2}.$$

(c) We can do it in two ways, either to verify that $\exp(-(1/2)kx^2/(k_B T))$ satisfies the Fokker-Planck equation with $\partial P/\partial t = 0$, or solve the equation $kxP + k_B T \partial P/\partial x = \text{const} = 0$ (the constant has to be 0 in order for $\int_{-\infty}^{+\infty} xP(x)dx$ finite).

9.7 (a) Ignore the transient term, the noise part of the solution is $v(t) = \int_{-\infty}^t e^{-\gamma(t-t')} R(t')/m \, dt'$. We can verify this is indeed the solution by substituting back to the equation. Position $x(t)$ is obtained by double integral over time. (b) Using the property of R correlation, we can do the integral with delta function, the result is $\langle R(t)x(t')\rangle = 2k_B T(1 - e^{-\gamma(t'-t)})\theta(t' - t)$. When $t' \le t$, we get 0. (c) Take the derivative of z, we get $dz/dt = d/dt\langle 2x\dot{x}\rangle = 2\langle \dot{x}^2\rangle + 2\langle x\ddot{x}\rangle$. Using the equipartition theorem for the first term since it is velocity square average and replacing the second derivative by the equation of motion, and using the result in (b), we get the required equation for z. Clearly, $dx^2/dt = 2xdx/dt$. And taking average and taking time derivative commute.

9.9 (a) We can derive the Fokker-Planck equation following the Zwanzig method, or we can just use the formal result in the textbook. The key point to note is that we have two independent variables x, and v, in addition to time t, since we are asking to find the joint probability of v and x. So, the equation of motion must be viewed as two coupled equations for x and v. Skip the math detail, the final Fokker-Planck equation is

$$\frac{\partial P}{\partial t} + \frac{\partial(vP)}{\partial x} + \frac{\partial\big[(-\omega^2 x - \gamma v)P\big]}{\partial v} = D\frac{\partial^2 P}{\partial v^2}.$$

Here we have defined $\omega^2 = k/m$, $\gamma = \gamma_L + \gamma_R$, and $D = k_B/m\,(\gamma_L T_L + \gamma_R T_R)$. (b) We assume that the Gibbs form is the solution, we are asked to determine the single parameter β. Setting the time derivative $\partial P/\partial t = 0$, and plugging in the trial form into the Fokker-Planck equation, we find the equation is satisfied provided that $\beta = \gamma/(mD)$, or stated differently, the effective temperature of the system to be a weighted average $T = (\gamma_L T_L + \gamma_R T_R)/(\gamma_L + \gamma_R)$ by the bath couplings. (c) To find the energy current, since we know in the steady state the probability distribution is Gibbsian with an effective temperature T, we can easily compute the velocity squared by applying equipartition theorem, but the velocity-noise correlation needs some thinking. So we can get quickly $J = \langle v(-m\gamma_L v + \xi_L) \rangle \rangle = -\gamma_L k_B T + \langle v\xi_L \rangle$. To find the later term, we need to go back to the Langevin equation to solve position or velocity (since $v = dx/dt$), we can focus on the position x. It is given by the Green's function of the system. The velocity at time t can be expressed as $v(t) = -1/m \int_{-\infty}^{t} \frac{\partial}{\partial t} G(t - t')\xi(t')dt'$, where the Green's function in frequency domain is given by $\tilde{G}(\omega) = \left[(\omega + i\eta)^2 - k/m + i\gamma\omega\right]^{-1}$. Using this result, the velocity and noise correlation can be computed, applying the noise correlation as delta function, we get $\langle v(0)\xi_L(0) \rangle = -2\gamma_L k_B T_L G'(0)$. Here the last term is the derivative of the Green's function at time $t \to 0^+$, which can be calculated using inverse Fourier transform and residue theorem, given $-1/2$. Thus, we have $\langle v\xi_L \rangle = \gamma_L k_B T_L$. Putting the two terms together, after some simplification, one obtains the desired result. It is possible to obtain the same result based solely on Fokker-Planck equation, see textbook in the supplementary reading.

10.1 Let Ω_0 be a set of phase space points at initial time $t = 0$. This region moves to Ω_t by time t. There is a one-to-one map from points $\gamma_0 \in \Omega_0$ to $\Gamma_t \in \Omega_t$. The Liouville theorem says (see, e.g., V.I. Arnold, "Mathematical Methods of Classical Mechanics," 2nd ed, page 69) $\int_{\Omega_0} d\gamma_0 = \int_{\Omega_t} d\Gamma_t$. [Is this also true when H is explicitly time dependent?] However, according to the usual rule of change of integration variables, we need

$$\int_{\Omega_t} d\Gamma_t = \int_{\Omega_0} \det\left(\frac{\partial \Gamma_t}{\partial \gamma_0}\right) d\gamma_0 = \int_{\Omega_0} D(t)\, d\gamma_0.$$

Comparing the two, we conclude that the Jacobian determinant $D(t)$ must be 1 for all t. Alternatively, $D(t)$ is the determinant of matrix elements (where Γ is at time t and γ_0 at time 0). Expand Γ for small time t using the Hamilton equations of motion, and taking the time derivative of the determinant and set time to 0, we obtain.

$$\frac{d\,D(t)}{dt}\bigg|_{t=0} = \sum_i^{2N} \frac{d}{dt}\left(\frac{\partial \Gamma_i}{\partial \gamma_i}\right)\bigg|_{t=0}.$$

Using the solution of Hamilton's equations near $t = 0$, e.g., $P_i = p_{0i} - t\partial H/(\partial q_{0i}) + O(t^2)$, we can show the sum above is zero due to the fact that the second order mixed derivative to H are equal, where $\Gamma = (\Gamma_1, \Gamma_2, \ldots, \Gamma_{2N})^T = (P_1, P_2, \ldots, P_N, Q_1, \ldots, Q_N)^T$ and similarly for the small γ version.

10.7 (a) The partition function is

$$Z = \frac{1}{h^3}\int_{-\infty}^{+\infty}\cdots\int_{-\infty}^{+\infty} dx\,dy\,dz\,dp_x\,dp_y\,dp_z\, e^{-\beta\left[\frac{p_x^2+p_y^2+p_z^2}{2m}+\frac{a}{2}(x^2+y^2+z^2)\right]}.$$

Here we note $r^2 = x^2 + y^2 + z^2$, and similarly $p^2 = p_x^2 + p_y^2 + p_z^2$. The calculation involves a 6-dimensional integral which can be done for each of the dimensions separately. We need to use the formula for Gaussian integration, which is $\int_{-\infty}^{+\infty} e^{-x^2/2}dx = \sqrt{2\pi}$. Using this result, we find $Z = [1/\hbar\beta\sqrt{m/a}]^3$. The free energy is $F = -1/\beta \ln Z$. The free energy difference is $\Delta F = F_f - F_i = 3/(2\beta)\ln(a_f/a_i)$, which is needed for part (b). (b) The Jarzynski equality is $\langle e^{-\beta w}\rangle = e^{-\beta(F_f - F_i)} = (a_i/a_f)^{3/2}$. The work done w is the difference of the Hamiltonians at the beginning and end of the process, $w = H(a_f, \mathbf{p}_f, \mathbf{r}_f) - H(a_i, \mathbf{p}_i, \mathbf{r}_i)$, where index i denotes the starting phase space point/parameter and index f the final ones. w can also be expressed as an integral over the path of the differential of the Hamiltonian.

10.8 (a) The Jarzynski equality takes the form $\langle e^{-\beta w}\rangle = e^{-\beta(F_B - F_A)}$, here $\beta = 1/(k_B T)$, and F_A is the free energy of the initial state and F_B is the free energy of the final state, $w = H_B - H_A$ is the microscopic work, and the average is over the initial canonical distribution $\rho_A = e^{-\beta H_A}/Z_A$. Although the expression for w looks simple, we must note that H_A is evaluated at the initial state (x_0, p_0) while H_B is evaluated at the final state $(x(t), p(t))$ which depends on the initial state and time t. The partition function at a fixed time t is $Z(t) = 1/h \int\int dx\,dp\, e^{-\beta(p^2/2m + k/2\,(x-vt)^2)}$. This is just a product of two gaussians one for p centered around 0 and another x centered around vt. The integral has the similar form, thus we obtain $Z(t) = 1/(\beta\hbar\omega)$, here we have defined the angular frequency as $\omega = \sqrt{k/m}$. Since the result is independent of time t, this means the free energy difference $F(t) - F(0) = 0$. (b) Applying the Hamilton equations of motion, we can eliminate the momentum, then the coordinate satisfies the equation $m\ddot{x} = -m\omega^2 x + vkt$. We can take care of the extra time dependent term if we use $x(t) = vt$. The full solution is a sum of sin and cos function together with the vt term. We can fix the initial condition by x_0 and p_0 of initial position and momentum. We find $x(t) = x_0 \cos\omega t + (p_0 - mv)/(m\omega) \sin\omega t + vt$. The momentum is obtained by differentiation, $p(t) = m\dot{x}(t)$. (c) The microscopic work w is completely determined by the initial position x_0, initial momentum p_0, and the time t, that is $w(x_0, p_0, t) = H(x(t), p(t), t) - H(x_0, p_0, 0)$. This expression can be simplified to $w = -2v\sin(\omega t/2)[m\omega x_0 \cos(\omega t/2) + (p_0 - mv)\sin(\omega t/2)]$. As far as the initial position and momentum is concerned, it is a linear function in them and t is just a fixed parameter. This expression is put into the Jarzynski exponential average of the work. (d) To evaluate the exponential work explicitly (not evoking the Jarnzynski theorem), we note the canonical distribution for the harmonic oscillator is just gaussian, and adding the extra $-\beta w$ term only shifts the center of the gaussian, after completing the squares we find it is just one, $\langle e^{-\beta w} \rangle = \int \int dx_0 dp_0 1/(hZ(0)) e^{-\beta(p_0^2/2m + k/2\, x_0^2 + w)} = 1$. (Steps omitted here). $Z(0) = 1/(\beta\hbar\omega)$ is the partition function.

11.1 $\sigma(\mathbf{p}\mathbf{p}_1|\mathbf{p}'\mathbf{p}_1') = \sigma(\theta, |\mathbf{p} - \mathbf{p}_1|)\delta^3(\mathbf{p} + \mathbf{p}_1 - \mathbf{p}' - \mathbf{p}_1')\delta(m(V^2 - V'^2)/2)$,

here $V = |\mathbf{p} - \mathbf{p}_1|$ and $V' = |\mathbf{p}' - \mathbf{p}_1'|$.

11.2 Using Liouville's theorem, $D\rho/Dt = \partial\rho/\partial t + \sum_j(\dot{q}_j\partial\rho/\partial q_j + \dot{p}_j\partial\rho/\partial p_j) = 0$, we have

$$\frac{dS_G}{dt} = -k_B \int \frac{\partial\rho\ln\rho}{\partial t}d\Gamma = -k_B \int (1 + \ln\rho)\frac{\partial\rho}{\partial t}d\Gamma$$

$$= k_B \int \left[\sum_j \dot{q}_j\frac{\partial\rho}{\partial q_j}(1 + \ln\rho) + \sum_j \dot{p}_j\frac{\partial\rho}{\partial p_j}(1 + \ln\rho)\right]d\Gamma$$

$$= k_B \int \left[\sum_j \frac{\partial(\dot{q}_j\rho\ln\rho)}{\partial q_j} + \sum_j \frac{\partial(\dot{p}_j\rho\ln\rho)}{\partial p_j}\right]d\Gamma = 0.$$

In the last step, we assumed that dq/dt is a function p only and dp/dt is a function of q only. Now the integrand is a divergence which can be converted into a surface integral using Gauss's theorem. We assume ρ is zero at far away surface, which gives us 0 for dS_G/dt.

11.3 The key steps are: $dH/dt = \int dr \int d\mathbf{p} \frac{\partial f}{\partial t}(1 + \ln f)$. This is because, t, $\mathbf{r}$, $\mathbf{p}$ are independent variables. Use Boltzmann equation $\partial f/\partial t = -\mathbf{p}/m \cdot \nabla_{\mathbf{r}} f - (f - f_{eq})/\tau$, the first momentum term can be dropped, because it can be written as $\int dr \nabla_{\mathbf{r}}(f \ln f)$ and can be changed to a surface integral using the Gauss theorem. The term 1 can also be dropped because particle number conservation, $\int dr \int d\mathbf{p} f = N$. We are left with $dH/dt = -\frac{1}{\tau} \int dr \int d\mathbf{p}(f - f_{eq}) \ln f$. To proceed further, we use the given local equilibrium condition, which means $\int dr \int d\mathbf{p}(f - f_{eq}) \ln f_{eq} = 0$. Dividing by $1/\tau$, adding into dH/dt, we get $dH/dt = -1/\tau \int dr \int d\mathbf{p}(f - f_{eq}) \ln(f/f_{eq}) \leq 0$.

Bibliography

Agarwalla, B. K., Li, B., and Wang, J.-S. (2012). Full-counting statistics of heat transport in harmonic junctions: transient, steady states, and fluctuation theorems, *Phys. Rev. E* **85**, p. 051142.

Amit, D. J. and Martin-Mayor, V. (2005). *Field Theory, the Renormalization Group, and Critical Phenomena*, 3rd edn. (World Scientific, Singapore).

Arfken, G. (1970). *Mathematical Methods for Physicists*, 2nd edn. (Academic Press, New York).

Arnold, V. I. (1989). *Mathematical Methods of Classical Mechanics*, 2nd edn. (Springer-Verlag, Berlin).

Ashcroft, N. W. and Mermin, N. D. (1976). *Solid State Physics* (Saunders College, Philadelphia).

Balescu, R. (1975). *Equilibrium and Nonequilibrium Statistical Mechanics* (John Wiley & Sons, New York).

Callen, H. B. (1985). *Thermodynamics and an Introduction to Thermostatistics*, 2nd edn. (John Wiley & Sons, New York).

Campisi, M., Hänggi, P., and Talkner, P. (1997). Colloquium: Quantum fluctuation relations: foundations and applications, *Rev. Mod. Phys.* **83**, p. 771.

de Gennes, P. G. (1966). *Superconductivity of Metals and Alloys* (CRC Press, Boca Raton).

Domb, C. (1996). *The Critical Point* (Taylor & Francis, London).

Eisenberg, D. and Kauzmann, W. (1969). *The Structure and Properties of Water* (Clarendon Press, Oxford).

Fermi, E. (1937). *Thermodynamics* (Prentice-Hall, New Jersey).

Ferrenberg, A. M., Xu, J., and Landau, D. P. (2018). Pushing the limits of Monte Carlo simulations for the three-dimensional Ising model, *Phys. Rev. E.* **97**, p. 043301.

Fetter, A. L. and Walecka, J. D. (1971). *Quantum Theory of Many-Particle Systems* (McGraw-Hill, New York).

Feynman, R. P. (1972). *Statistical Mechanics: A Set of Lectures* (Addison-Wesley, Reading).

Franzosi, R. (2018). Microcanonical entropy for classical systems, *Physica A* **494**, p. 302.

Friedman, H. L. (1985). *A Course in Statistical Mechanics* (Prentice-Hall, Englewood Cliffs, New Jersey).

Georges, A., Kotliar, G., Krauth, W., and Rozenberg, M. J. (1996). Dynamical mean-field theory of strongly correlated fermion systems and the limit of infinite dimensions, *Rev. Mod. Phys.* **68**, p. 13.

Gibbs, J. W. (1902). *Elementary Principles in Statistical Mechanics* (Yale University).

Goldstein, H., Poole, C., and Safko, J. (2002). *Classical Mechanics*, 3rd edn. (Pearson Education Inc., San Francisco).

Grimus, W. (2011). On the 100th anniversary of the Sackur-Tetrode equation, https://arxiv.org/abs/1112.3748.

Gubernatis, J., Kawashima, N., and Werner, P. (2016). *Quantum Monte Carlo Methods: algorithms for lattice models* (Cambridge University Press, Cambridge).

Hansen, J.-P. and I. R. McDonald (2006). *Theory of Simple Liquids*, 3rd edn. (Academic Press).

Harris, S. (2004). *An Introduction to the Theory of Boltzmann equation* (Dover, Mineola, New York).

Haug, H. and Jauho, A.-P. (1996). *Quantum Kinetics in Transport and Optics of Semi-Conductors* (Springer-Verlag, Berlin).

Huang, K. (1987). *Statistical Mechanics*, 2nd edn. (John Wiley & Sons, New York).

Isihara, A. (1971). *Statistical Physics* (Academic Press, New York).

Jarzynski, C. (1997). Nonequilibrium equality for free energy differences, *Phys. Rev. Lett.* **78**, p. 2690.

Kalos, M. H. and Whitlock, P. A. (1986). *Monte Carlo Methods, Vol. I: Basics* (John Wiley & Sons, New York).

Khinchin, A. I. (1949). *Mathematical Foundations of Statistical Mechanics* (Dover, New York).

Kubo, R. (1965). *Statistical Mechanics: An Advanced Course with Problems and Solutions* (North-Holland Physics Publishing, Amsterdam).

Kubo, R., Toda, M., and Hashitsume, N. (1992). *Statistical Physics II, nonequilibrium statistical mechanics* (Springer, Heidelberg).

Landau, D. P. and Binder, K. (2015). *A Guide to Monte Carlo Simulations in Statistical Physics*, 4th edn. (Cambridge University Press, Cambridge).

Landau, L. D. and Lifshitz, E. M. (1965). *Quantum Mechanics: non-relativistic theory*, 2nd edn. (Pergamon Press).

Langevin, P. (1908). On the theory of Brownian motion, *C. R. Acad. Sci (Paris)* **146**, p. 530, english translation by D. S. Lemons and A. Gythiel, *Am. J. Phys.* **65**, 1079 (1997).

Lebowitz, J. L. (1974). Exact derivation of the van der Waals equation, *Physica* **73**, p. 48.

Lebowitz, J. L. (2007). From time-symmetric microscopic dynamics to time-asymmetric macroscopic behavior: an overview, https://arxiv.org/abs/0709.0724.

Lieb, E. H. and Yngvason, J. (1999). The physics and mathematics of the second law of thermodynamics, *Physics Reports* **310**, pp. 1–96.

Lindenberg, K. and West, B. J. (1990). *The Nonequilibrium Statistical Mechanics of Open and Closed Systems* (Wiley-VHC, New York).

Liu, S., Agarwalla, B. K., Wang, J.-S., and Li, B. (2013). Classical heat transport in anharmonic molecular junctions: exact solutions, *Phys. Rev. E* **87**, p. 022122.

Ma, S.-K. (1976). *Modern Theory of Critical Phenomena* (Benjamin/Cummings, Reading).

Mahan, G. D. (2000). *Many-Particle Physics*, 3rd edn. (Kluwer Academic, New York).

Maxwell, J. C. (1888). *Theory of Heat*, ninth edn. (Longmans, Green and Co./Dover 2001, London).

Mayer, J. E. and Mayer, M. G. (1975). *Statistical Mechanics*, 2nd edn. (Wiley).

McQuarrie, D. A. (2000). *Statistical Mechanics* (Mill Valley, California).

Metropolis, N., Rosenbluth, A. W., Rosenbluth, M. N., Teller, A. H., and Teller, E. (1953). Equation of state calculations by fast computing machines, *J. Chem. Phys.* **21**, p. 1087.

Norris, J. R. (1997). *Markov Chains* (Cambridge University Press, Cambridge).

Onsager, L. (1944). Crystal statistics. I. A two-dimensional model with an order-disorder transition, *Phys. Rev.* **65**, p. 117.

Pathria, R. K. (1972). *Statistical Mechanics* (Pergamon Press, Oxford).

Pippard, A. B. (1957). *The Elements of Classical Thermodynamics* (Cambridge University Press, Cambridge).

Planck, M. (1914). *The Theory of Heat Radiation*, 2nd edn. (P. Blackiston's Son & Co., Philadelphia), translated into English by M. Masius.

Plischke, M. and Bergersen, B. (2006). *Equilibrium Statistical Physics*, 3rd edn. (World Scientific, Singapore).

Pottier, N. (2010). *Nonequilibrium Statistical Physics* (Oxford University Press, Oxford).

Press, W. H., Teukolsky, S. A., Vetterling, W. T., and Flannery, B. P. (1992). *Numerical Recipes in C*, 2nd edn. (Cambridge University Press, Cambridge).

Reichl, L. E. (1980). *A Modern Course in Statistical Physics* (Edward Arnold, London).

Reif, F. (1965). *Fundamentals of Statistical and Thermal Physics* (McGraw-Hill, Auckland).

Rieder, Z., Lebowitz, J. L., and Lieb, E. (1967). Properties of a harmonic crystal in a stationary nonequilibrium state, *J. Math. Phys.* **8**, p. 1073.

Risken, H. (1989). *The Fokker-Planck Equations*, 2nd edn. (Springer, Berlin).

Rubin, W. (1976). *Principles of Mathematical Analysis*, 3rd edn. (McGraw-Hill Book Co., Singapore).

Schultz, T. D., Mattis, D. C., and Lieb, E. (1964). Two-dimensional Ising model as a soluble problem of many fermions, *Rev. Mod. Phys.* **36**, p. 856.

Sharp, K. and Matschinsky, F. (2015). Translation of Ludwig Boltzmann's paper on the relationship between the second fundamental theorem of the mechanical theory of heat and probability calculations regarding the conditions for thermal equilibrium, Sitzungberichte der Kaiserlichen Akademie der Wissenschaften. Mathematisch-Naturwissen Classe. abt. ii, lxxvi 1877, pp. 373–435 (Wien. Ber. 1877, 76:373-435). reprinted in Wiss. Abhandlungen, vol. ii, reprint 42, p. 164-223, barth, leipzig, 1909, *Entropy* **17**, 4, pp. 1971–2009.

Smith, H. and Jensen, H. H. (1989). *Transport Phenomena* (Oxford University Press, Oxford).

Stanley, H. E. (1971). *Introduction to Phase Transitions and Critical Phenomena* (Oxford University Press, Oxford).

Stauffer, D. and Aharony, A. (1994). *Introduction to Percolation Theory*, 2nd edn. (CRC Press, Boca Raton).

Suzuki, M. (1995). *Coherent-Anomaly Method: Mean Field, Fluctuations and Systematics* (World Scientific, Singapore).

Swendsen, R. H. (2012). *An Introduction to Statistical Mechanics and Thermodynamics* (Oxford, London).

Thompson, C. J. (1988). *Classical Equilibrium Statistical Mechanics* (Oxford University Press, Oxford).

Toda, M., Kubo, R., and Saitô, N. (1992). *Statistical Physics I*, 2nd edn. (Springer, Berlin).

Vasil'ev, A. N. (2004). *The Field Theoretic Renormalization Group in Critical Behavior Theory and Stochastic Dynamics* (Chapman & Hall/CRC, Boca Raton).

von Neumann, J. (1955). *Mathematical Foundations of Quantum Mechanics* (Princeton Univ Press).

Wang, J.-S., Ni, X., and Jiang, J.-W. (2009). Molecular dynamics with quantum heat baths: application to nanoribbons and nanotubes, *Phys. Rev. B* **80**, p. 224302.

Yeomans, J. M. (1992). *Statistical Mechanics of Phase Transitions* (Oxford University Press, Oxford).

Ziman, J. M. (1960). *Electrons and Phonons* (Clarendon Press, Oxford).

Zwanzig, R. (2001). *Nonequilibrium Statistical Mechanics* (Oxford University Press, Oxford).

Index

Printed in the United States
by Baker & Taylor Publisher Services